Private Television Communications: The New Directions

The Fourth Brush Report

by
Judith M. and Douglas P. Brush

H I Press of Cold Spring, Inc.
Cold Spring NY 10516

in association with the
International Television Association

384.5543 B912pn
Brush, Judith M.
Private television
communications

**PRIVATE TELEVISION COMMUNICATIONS:
THE NEW DIRECTIONS**

Copyright © 1986 by Judith M. and Douglas P. Brush. All rights reserved. Printed in the United States of America. No part of this book may be used or reproduced in any manner whatsoever without the written permission, except in the case of brief quotations enbodied in published articles and reviews. All inquiries should be addressed to: H I Press, Inc., P.O. Box 361, Cold Spring, NY 10516.

First Edition

Cover Design by Joseph D. Kowal

Portions of this book have appeared in article form as the "Brush Report" at various times in *Video Pro* magazine.

Library Congress Catalog Card Number: 86-080423
ISBN: 0-9615988-0-8

To those who waited so patiently for the study to be completed.

Acknowledgments

First and foremost in our acknowledgments is a "Thank You!" to the private television users who took the time to answer our lengthy questionnaire. Without their faithful participation in "The Brush Reports," our readers would not have the information they need to make their plans.

We also would like to thank the Board of Directors of ITVA under the leadership of Kathy Morris, its President, Steve Mulligan, its Executive Director, Fred Wehrli and attorney Mike Reed for their patience and assistance in bringing this project to fruition. And we also thank Lynn Yeazel, David Smith and Rita Marie Sansone for their valuable input and comments during the final edit.

And finally we wish to thank our son, Douglas Alexander for giving up time with his parents so they could get the book to press.

Judith and Douglas Brush

Cold Spring-on-Hudson, NY
January 1986

CONTENTS

Foreword

"Beam Me Up, Scotty!"   1
By Steve Mulligan, President, International Television Association

Introduction

Why Another "Brush Report"   5
Scope of the Study   7
The Survey   9
How the Data Was Processed   11
Data Interpretation   11

Chapter 1 — Setting the Stage

How It All Began   13
What Video Does   13
Early Applications   15
The Awakening Giant   17
Into the Eighties   17

Chapter 2 — Stone Tablets to Storyboards

All Communications Is Essentially Visual   19
Enter Media Technology   22
The Picture Changes   24
Adrift in the Data Stream   25
Pumping Video   26

Chapter 3 — How Organizations Communicate

The Rumor Mill	31
Credibility	31
Perceptions of Video	32

Chapter 4 — Determining Communications Needs

Video Is Not A Fad	35
Communications Needs Analysis	36
Communications Task Force	37
Information Channels	38
Information Categories	39
Information Shelf-Life	40
Information Flow	41

Chapter 5 — Management and The Tube

Proposals to Management	46

Chapter 6 — Who Is Using Private Television?

Video Is Everywhere!	51
Users By Size and Type of Organization	52
Typical Video User	54
Years of Private Television Experience	55

Chapter 7 — Video Uses and Applications

Just Another Tool	58
Training vs. Communications	60
Applications In Ranking Order	62
Future Uses of Video	64

Chapter 8 — Programming

Monkey Island	67
The Other End of the Horse	69
Levels of Production	71
Video Memos	74
Program Sources	74
Published Programs	75
Length of Programs	75
In-House Clients for Video Programming	75

Chapter 9 — Production

Investment in In-House Production Equipment	78
Investment in Post-Production Equipment	80
Location Shooting	80
Size of In-House Studio or Production Area	80
Location of In-House Studio or Production Area	82
Production Formats	83
Mastering Formats	84
The New 8 MM Format	84
Post-Production Equipment	85
Supplemental Outside Services	85
In-House Studios vs. Outside Production Services	86

Chapter 10 — Distribution

Are You Suffering From "Network Narcosis?"	93
What is "Network Narcosis"?	94
Viewing Locations	99
Video Networks	99
Network Expansion	101
Selective Distribution	101
International Distribution	102
Languages In Which Programs Are Distributed	103
How Private Television Programs Are Viewed	103
Number of Times Programs Are Played	104
Means of Program Distribution	104
Distribution Equipment Investment	106
Reuse of Videotapes	108

Chapter 11 — Promotion, Evaluation and Feedback

Program Promotion	109
Validation and Measurement of Audience Reaction	110
Measurements Used	111

Chapter 12 — Budgets, Charge Backs and Cost Analysis

Size of Video Budgets	113
Budget Expenditures	114
Cost Per Program	114
New Equipment Purchases	116
Charge-Back Systems	117
Cost Analysis	120
Fixed, Operating and Production Costs	121
Cost Per Project	122
Project Estimating	123
The Bottom Line	123

Chapter 13 — Organization and Staffing

Where Video Fits In The Organization	125
Who's In Charge	126
Reporting Level	128
Video Centralization	128
End Users As Producers	129
Cooperation With Other Video Units	130
Video Policy	130
Video Staffing	130
Job Titles	131

Chapter 14 — Interactive Video

Interactive Video Is Growing	137
Tape vs. Disc	137
Numbers and Applications of Interactive Programs	138
Locations Using Interactive Video	138
Video Manager's Involvement In Interactive Video	139

Chapter 15 — The New Technologies

Local Area Networks	141
Computers in the Video Department	143
Computer Graphics	144

Chapter 16 — Satellite Videoconferencing

Systems In Use	148
Numbers of Videoconferences and Locations	150
Primarily a Management Medium	151

Chapter 17 — The Market

Size of the Industry	154

Appendices

A—Questionnaire	157
B—Job Descriptions	169
C—Bibliography	185
D—Publications	187

List of Tables

Introduction

I-1 Private Television Respondents by Primary Business/ Organization Classification — 10

Management and The Tube

5-1 Management Level At Which Decision Was Made To Use Private Television — 45
5-2 Appearances By Head of Organization — 46

Who Is Using Private Television?

6-1 Private Television Respondents by Number of Employees/Members — 53
6-2 Years of Private Television Experience — 56

Video Uses and Applications

7-1 Current Video Applications — 63
7-2 Current Video Applications Compared With Uses in 1981 — 64
7-3 Future Video Applications in Ranking Order of Probable Use — 65

Programming

8-1	Median Number of Programs Produced	72
8-2	Number of Private Television Programs Produced in 1985	73
8-3	Average Running Length of Programs Produced	76
8-4	Organizational Unit For Whom Video Unit Produces Programs	76

Production

9-1	Source of Private Television Programs	77
9-2	Total Investment In In-House Production Equipment	79
9-3	Total Investment In In-House Post Production and Editing Equipment	81
9-4	Size of In-House Studio or Fixed Production Area	82
9-5	Location of Studio or Fixed Production Area	83
9-6	Videotape Production Formats	83
9-7	Supplemental Outside Services Used	86

Video Network and Distribution Systems

10-1	Number of Viewing Locations	100
10-2	Growth Rate of Viewing Locations	101
10-3	Summary of Number of Locations	101
10-4	Average Number of Locations to Which Programs Are Distributed	102
10-5	How Private Television Programs Are Viewed	104
10-6	Means of Private Television Program Distribution	105
10-7	Primary Means of Program Distribution	105
10-8	Total Investment In Distribution/Playback Equipment.	107

Program Promotion and Evaluation

11-1	Forms of Video Program Promotion Used to Stimulate Viewing	110
11-2	Private Television Users Validating/Measuring Program Results	111
11-3	How Audience Reaction Is Validated/Measured	112

Budgets, Charge Backs and Cost Analysis

12-1	1985 Operating Budgets for Private Television	113
12-2	Reported Average Cost Per Program	115
12-3	1985 Expenditures For New Equipment	116
12-4	Private Television Users Having Charge-Back Systems	118

Organization and Staffing

13-1	Administration of Video Operation	127
13-2	Type of Video Policy In Effect	130
13-3	Number of Employees Devoting All or Major Portion of Time to Video	131

Interactive Video

14-1	Media Used For Interactive Video Applications	138

Foreword
"Beam Me Up, Scotty!

*by
Steve Mulligan, President
International Television Association*

Have you ever wondered if there is life on other planets? Scientists tell us that each of the millions of galaxies we can see contain millions of suns much like ours. Statistically, many of these can support a solar system like ours and perhaps even a planet like earth capable of sustaining life like ours. However, the possibility of life on that planet being *exactly* at the same point in development as ours is very remote. It is more probable that the life on any other planet is either somewhere behind us or very much ahead of us.

Development is a result of a number of things, such as . . .

- when the planet was created;
- the environmental factors that would affect life;
- how life on that planet would cope with changes;
- the influence its leadership had on life there.

If there is life on other planets, each of these factors will influence their state of development in relation to ours.

If you ever watched the great television series "Star Trek", you saw the characters fall into situations where the life on other planets was either much behind or ahead of that on the Starship Enterprise. Some were making the same mistakes others made centuries before. Others never made many mistakes and had moved ahead rapidly. Still others seemed to be stuck, for whatever reason, at a certain point in development. In all cases, development was taking place at varying speeds, no matter where they were on the continuum. Life was going on and progress of some sort was taking place. No one was wrong for being in their particular part of the continuum. They simply were where they were.

What does this have to do with organizational television? Quite a bit, actually. Consider for a moment that each video facility or video business, whether attached to an organization or standing alone, whether one person (including independents) or many, is unto itself a separate planet. Each of us at one time or another has wondered if there is "life as we know it" in other video operations. Sometimes we wonder if there is anyone at all out there, or whether we are operating in a vacuum, forever doomed to keep reinventing the wheel.

Do others face the same frustrations? Do others have the same technical problems or management inhibitions? Who doesn't wish that they could get Scotty to "beam" them to another corporate video operation to see where they are in terms of development and to learn how they may have handled the problems we may be facing currently.

If we were to be able "to explore strange new worlds . . . to seek out new life and new civilizations . . . to boldly go where no man has gone before" in the video universe, we would find that some are way ahead of us and others are far behind. We would probably feel good about being ahead of some but feel challenged about being behind the others. We would offer suggestions to those who are following us, and learn from those who lead the way. We would conclude that there is a continuum that everyone seems to be following . . . a continuum of development.

The degree and speed of progress and development would vary from place to place. All the conditions that affect the development of life on a planet — environment, coping, creation, decisions, leadership and so on — affect the progress of the video operations. Each are at various points on the continuum for different reasons.

My year as President of ITVA has been like an interstellar voyage on the Enterprise. I have been "beamed" into many facilities and have talked with hundreds of video professionals. This experience has enabled me to see video "life as we know it" in many different stages of development on many different video "planets." Here are some of the things I have learned.

Some of the facilities and video businesses developed like the dinosaurs and died off. Changes in the technical and economic environment saw to that. Like the phoenix, however, many have risen from the ashes to become stronger and more powerful resources to the organizations they serve.

Foreword 3

Over all, I see a speeding up of progress. The environment has changed in many places. The economic forces that "leaned up" the American corporations and organizations also have created a new respect for the value of organizational television. While video can be a cost saver, more importantly it can take limited resources (the new lean management staff) and spread them consistently to a large number of people.

The leadership of the video businesses and facilities have learned a new entrepreneurship. They are more creative in proving video as a resource. They are matching video technology to the bottom line and spreading the word throughout the organization. They are better trained and better informed and are helping their clients make smarter decisions.

Those who are just beginning are starting out on a sounder footing. The need for video for communications and training is stronger now than it was five years ago, when many saw technology as a *nice to have* rather than a *necessity*. The need to communicate in this information age is what is behind the start up of many new video operations.

Right now I see the whole process of development speeding up for everyone. For some it is growing at exponential proportions. One success builds two more, then four, then sixteen and so on. More and more media managers are finding that their advice is being heeded in the board rooms. Their expertise in communications problem solving is being actively sought. In many organizations today, the successful introduction of any new corporate program or policy frequently hinges on the video program that introduces and positions it.

It is really exciting to see what is happening out there on all those creative video planets. The momentum is growing stronger every day. If you are just starting or are on the first rungs of the ladder, take heart and be ready. Video professionals everywhere are reaching for those stars and getting there. Be assured that life on other planets is alive and well and is developing steadily.

You are not alone!

Introduction

Why Another "Brush Report"

Thirteen years ago we helped organize the first national conference of the newly formed International Television Association (ITVA) at the National Association of Broadcasters Convention (NAB) in Washington DC. The ITVA had just been formed by the merger of two separate video user organizations, the Industrial Television Society (ITS) and the National Industrial Television Association (NITA).

The opening session was the inauguration of a new television transmission system in Washington called Multipoint Distribution Service (MDS). This inaugural program featured ITVA's first president, Lynn Yeazel (currently a member of ITVA's board), FCC Chairman Dean Burch and White House Communications Director Herbert Klein. The term "private television" was coined at this meeting with the new transmission service being described as "making possible hundreds of *private television* networks."

The term "private television" seemed to aptly describe our then unnamed industry made up of only a few hundred organizations which were using television for communications and training. The term was broader in scope than "industrial television", more positive than "non-broadcast" and clearly superceded "closed-circuit" as a term applied to the fast growing area of corporate communications. The new name stuck and has been used ever since to describe the industry as a whole in this country and abroad.

When our first study, *"Private Television Communications: A Report to Management"* was published by Knowledge Industry

Publications in 1974, it was to be a one-shot effort. Our purpose was to establish "for the first time the true dimensions of the rapidly expanding non-broadcast television industry." Then we were going on to other things. The "other things" are still waiting.

At that time we saw from first hand experience that video was fast being adopted by hundreds of business, government and non-profit organizations. We also saw that its use was beginning to have a major impact on these organizations by revolutionizing management and employee communications, personnel and training practices and marketing and sales activities.

Our purpose then, as it remains today, was to investigate the wide range of uses and applications of private television so that the management of both present and future video user organizations could make more effective decisions on their own use of this vital new communications tool.

In addition, we tried to develop a body of data which would enable the manufacturers and suppliers serving the industry to be more understanding of how the medium is being used. Our goals for the present study — now 13 years later — remain the same.

In 1974, we reported that there were more than 300 corporations, and other organizations spending over $48.5 million for organizational television program production and distribution. These "user/producers", as we still call them, originated more than 13,000 individual productions totalling over 3,500 hours. This was more programming than was carried in prime time by all three commercial television networks combined.

At that time we identified some 75 private video "networks". We defined a "network" as any video distribution system carrying user-oriented program material, either in-house or custom-produced, to six or more locations away from the point of origination on a regular basis.

The total size of the market then — including both hardware and software — was $207 million and was growing at a rate in excess of 30 percent annually.

Since then, the actual growth of the private television industry has exceeded our most optimistic predictions. Today we estimate that some 8,500 organizations are producing some form of video for communications and/or training. Last year this resulted in approximately 55,000

Introduction 7

hours of programming, now a figure too large to be related meaningfully to anything.

The reasons behind this growth — some of it occurring during two very intense economic recessions — are as varied as the medium's applications. There are, however, two fundamental underlying factors: *first*, the very real communications and training needs which private television seems to be meeting more effectively than any other medium; *second*, the increasing availability of low-cost, easy-to-use equipment, coupled with imaginative objective-oriented programming concepts and production techniques which make video a cost-effective practicality for most organizations.

Little did we realize 13 years ago that our "one shot effort" would become a lifetime commitment and that we would be working on a "Fourth Brush Report" in the mid-Eighties. When each of the preceding studies was finished, we vowed it would be the last. Then, three years later, the phone would start ringing with requests urging us to do another.

Well, here it is.

Scope of the Study

Although a "Brush Report" is published approximately every four years, the research which produces the report is an ongoing and ever-evolving activity.

The major source of our information for each "Report" comes from our day-to-day contact with video users, with professional video production houses and individual producers, with suppliers of services and manufacturers of products aimed at the organizational communicator and trainer, and with educators who are preparing students for employment in the field.

Our routine consulting work provides us with ample opportunity to observe the functioning of the video and audiovisual departments of diverse corporations in many different industries. We see how top management's support, or lack of it, influences the role of the visual communicator in his or her organization. We hear the problems which may arise for the visual communicator when his or her organization adopts yet another management fad.

We also have the opportunity to work with senior management and Human Resources departments in determining the job titles, career pathing and salary levels of video and audiovisual people.

Through field interviews and focus groups with everyone from assembly line workers to office personnel to divisional or regional management, we find out what they, as receivers of the multitudinous messages in all media, think about organizational communications and training. Essentially they tell us what works and what doesn't, what media are suited for the workplace and which have no business even being introduced into their environment.

These observations and experiences help us to spot what we feel are trends in the use of video and other communications media. All of this information is synthesized into the "Brush Reports". Our contacts and observations have led to the revision and polishing of the mail survey questionnaire used to gather the hard data for our reports. This study's questionnaire is the longest to date. It asks the video user to answer 161 questions, some of them complicated and multilayered. The 1973 study, on the other hand, asked only 29 questions, covering only the basics like applications, budgets, type of distribution systems and size of staff. The 1977 questionnaire grew to 47 questions and the 1981 to 67.

This year's questionnaire reflects the more complicated nature of the private television industry and the proliferation of new products and media which are influencing the future of video as a whole.

The questionnaire is divided into two parts. The main body consists of 128 questions covering budgets, administration, programming applications and production, distribution and networking, evaluation and promotion, video projection, interactive video and professional education and training. The second part is a special section on videoconferencing consisting of 33 questions. (Appendix A)

Not all of those who received the questionnaire were able to complete the entire 161 questions because certain areas did not apply to their video operation or to their overall organization. Very few of the respondents complained about the length of the questionnaire and, once again as has happened in the past, many video managers used the questionnaire as a basis for their own yearly planning activities.

In doing the first three studies we developed a "qualified" list of video user organizations to whom we have sent our questionnaires. We consider this a "qualified" list because these are organizations who originate their own video programs on a regular basis for communications and training. A large number of these are ITVA members and

Introduction

have responded to our past surveys as well as have participated in phone or personal interviews or in focus groups over the years. These are organizations that we feel we can use to "track" the growth of the industry.

Once again as in the past, we have excluded certain types of organizations from our report: organizations which merely buy or lease published programs for training purposes but do not produce any programs of their own. Nor do we include the users of video as an instructional device, such as schools and universities (ETV), or applications where video is used purely as an archival device to record and preserve an activity or event. These are highly specialized applications and are covered in many of the fine professional and trade journals devoted to those audiences.

And, we have not included closed-circuit television applications in the report since CCTV is basically a monitoring or recording activity and not a communications or training application. CCTV utilizes different kinds of equipment which are separated by manufacturers and dealers from the hardware catalogues and trade show booths aimed at the private television user. Since our first study, the term CCTV has generally fallen into disuse in relation to communications and training applications.

Home video also will not be covered *per se* in this report. It will be mentioned to some extent since the lines between home and private video are more blurred than ever before.

The Survey

Our survey mailing list consisted of three parts. The first segment is our "qualified video user" list. This time it was more difficult to update our list than ever before. With *mergermania* and *name-changeitis* (1985 was the second most active year in history for corporate name changes) running amok in our boardrooms, it was a real challenge to find some of our former questionnaire respondents. A handful of companies have literally vanished off the face of the corporate earth.

The second part of the sample was selected from a carefully-screened group of organizations fitting the basic video user demographic profile which we have developed through the first three studies. The video usage of many of these organizations was not known.

The third part of the sample is made up of a representative selection of non-profit organizations and government agencies. These are all known video users and all happened to be ITVA members as well.

The total sample size for the mail survey was 1,100 organizations. We received a response from 248 organizations, a return of 22.6 percent.

As Table I-1 indicates, the response of those in the non-industrial, or service sector was higher with 56.1 percent of the users in that category while those in the manufacturing sector represented 34.7 percent of the user responses.

TABLE I-1
Survey Respondents by Primary Business/Organization Classification
N = 248

Classification		Total Respondents
Manufacturers	(86)	34.7%
Non-Industrials	(139)	56.1
Medical/Education	(9)	3.6
Government	(9)	3.6
Other	(5)	2.0
	(248)	100.0%

In the 1981 survey, manufacturers made up 37.1 percent of the respondents while non-industrials represented 41.7 percent. In 1977, the two groups were evenly divided, while in the first study in 1973, 53 percent of the respondents were manufacturers and only 28 percent were non-industrials.

The current study, therefore, reflects the shift in our economy from a manufacturing to a non-industrial or service industry base which began to show its impact in the mid-Seventies.

Looking more closely at the non-industrial sector respondents, 19 percent are insurance companies and 32 percent are utility companies. Both industries are service oriented with consumer as well as employee information of prime importance.

The remainder of the respondents were medical, educational, non-profit and government agencies. They represent a total of 9.2 percent of the respondents.

How the Data Was Processed

While reporting the changes in technology, we were also affected by them. Survey data from the last two studies was processed on an IBM 370 mainframe computer using the SSPS statistical analysis program. This involved knowing exactly what you wanted to ask before putting the data through the computer. Preliminary runs often raised new questions and the entire job would be run again to answer them.

This time the data was processed by a standard desk top PC computer that came close to rivaling the capability of the big mainframe. However we now had the capability of asking the computer any question we could think of in real time on a query basis without having to run an entire program. In addition, we can continue to update the data on a continuing basis and respond to very specific inquiries more readily than before.

Data Interpretation

We repeat the data interpretation caveat that we have made in each of the preceding studies. It is unfortunate that many video managers are still not up to speed on cost accounting and reporting procedures. Thus, their estimates of average program costs, for instance, were below the *actual* dollar figure in spite of the fact that we have asked them to include *all* costs in their cost-per-program figure. In addition, in determining annual operating expenditures many did not include salaries and overhead.

In these instances, we have interpreted the results with our own experience base and have attempted to present a picture closer to reality than the submitted data would otherwise provide.

Chapter 1
Setting the Stage

How It All Began

It is easy to forget just how recent a phenomenon television is and how radically it has changed our lives, especially since more than half of our population has never known a world *without* television.

In 1938 (a year within the living memory of some of us) RCA's founder, David Sarnoff, was laughed out of the room when he told a meeting of the Radio Manufacturers Association that, "television in the home is now technically feasible." A leading trade publication of the time even went so far as to use Sarnoff's remarks as the basis for a promotional mailing which included a century plant seed and the prediction that when it first bloomed, then would be the time that television would be a reality.

Actually, television in an embryonic stage, had been around sometime before that. The first recorded video image was on an Edison-style gramaphone *disc* in England in 1927. The inventor, John Logie Baird, called it "Phonovision" but saw little practical use for it since, as he said: "the cinematograph (motion picture) serves the same purpose, and does it in a much more perfect fashion." Nevertheless, a few of Baird's videodiscs were sold commercially in 1935 for $1.00 each. They would be worth a fortune to collectors today.

What Video Does

Critics have often chided television as an imperfect substitution for something else. Does it, in fact, possess characteristics unique onto itself that no other medium can offer? The answer has always been yes and no.

In truth, for many applications, video *is* just another audiovisual device, and, as such must co-exist with a wide range of other media. But television is also — as its success as a mass medium has abundantly demonstrated — something different.

There is something in the nature of the video image that no one really understands but which is universally acknowledged. And that is <u>people will watch a television screen no matter what is going on around them</u>.

In 1963, psychologists observing the effects of the televised coverage of John F. Kennedy's funeral were astounded to note that spectators lining the route of the motorcade stared doggedly at the TV monitors placed along the way even though the actual event was taking place only a few feet from them. Even seasoned print journalists found themselves reporting the story as it appeared on television in spite of the fact that they had front-row viewing locations.

Today, you will find fans in the stands at baseball and football games watching the game on small portable TV sets. They will tell you that they can see the plays more clearly on televison and have a better understanding of the action. Some will even say that it is more entertaining to watch the game on television even though they are surrounded by thousands of screaming spectators.

In our own research, we conducted a number of informal experiments with a simple, single-camera, recorder and monitor to try to evaluate the impact of the television image. In an ordinary social situation we set up the camera and focused on one of the more attractive ladies in the room. Her image could be seen on a monitor directly across from where she was sitting. As we sat and talked everyone related to her *on the monitor* rather than in person. As the experiment continued, she became increasingly annoyed that people would only talk to her video image instead of her directly. When we played back the tape, she still could not compete for direct attention with her own video persona!

Some experts have said that television is the most effective form of communications ever devised. There seems to be something hypnotic about a television picture. As a matter of fact, hypnotists report that they can put a subject into a trance more easily using video than they can face-to-face.

Perhaps it is the constantly moving, kinetic effect of the scanning lines. A video image is never static, even when showing a still photo. Unlike a motion picture which is composed of a rapid series of static or frozen images, the television image is *alive*, a constantly fluid kaleidoscope of motion and change.

Setting the Stage 15

Perhaps it is due also to the fact that technically no image ever really appears on the screen. There is only a rapidly moving dot of light of varying color and intensities which the human brain converts into a picture. With film, the images are always there and the brain has to create only the motion not the picture itself. In this sense, video is by nature a more interactive medium since it requires the participation of the viewer in order to complete the process.

It will be a long time before the full psychological, philosophical, sociological and anthropological effects of this medium are ever fully understood.

Early Applications

While the extensive use of private television is less than 20 years old — with most of its development taking place during the last 10 years — many organizations had been using it on an infrequent basis for a much longer period. Almost from the time that television became a commercial medium, various companies began finding ways of harnessing its power and impact for private communications purposes.

Prior to the development of the broadcast videotape recorder in the mid-1950's, video programs had to be presented live over internal closed-circuit systems or sent out over telephone company lines to remote locations. This was a very costly undertaking. Broadcast TV equipment had to be used and telephone company video lines were very expensive. Because of the high cost, such presentations had to be made to fairly large audiences which were forced to watch the programs on relatively small monitors or on dim video projection systems.

The first interactive use of the medium, which employed a two-way audio circuit, was in the early 1950's when the GE Small Appliance Division held a live, nationwide sales meeting to get dealer reaction to a critical inventory problem.

Following management's presentation of the problem, dealers in each city got a chance to speak their minds directly. Finally one dealer stood up and bluntly suggested that the company cut prices on the old inventory and move on to the new product line. He was cheered from coast-to-coast and the company knew — decidedly — what action to take.

The technology involved in this first interactive videoconference is also interesting. The dealer audience in New York was too large for the use of monitors, and no effective large-screen video projection systems were then in existence.

The meeting was held in a large New York movie theater. On the floor above the projection booth a special movie camera was set up in front of a small TV screen. The exposed film was fed directly from the camera into a fast film processor and then, still wet, through a hole in the floor and into the movie projector below for showing to the audience.

The need for live telecasts, and the attendent high transmission costs, obviously limited the use of television for private communications until the mid-Sixties when the first helical scan, black-and-white videotape recorders were introduced. The first institutional or industrial units were one-inch systems — either IVC or Ampex — which were not compatible with each other.

These were followed on the market by a series of 1/2-inch open-reel portable units made by several different manufacturers, none of which were interchangeable with each other. Aimed initially at the consumer market, the half-inch units quickly found homes in corporate and institutional training departments then in desperate need of fast, low-cost, do-it-yourself media for data processing training.

The one-inch systems were quickly adopted by the larger audiovisual and training departments where they were installed in studios complete with cameras, switchers, and other production paraphernalia. Less affluent organizations bought the half-inch home units and a single camera to turn out "fast and dirty" training tapes, often of nothing more than someone lecturing with a flip-chart or chalk board.

Dreadful by today's standards, these tapes nevertheless were an instant hit because most viewers were vitally interested in the content. In addition, the medium carried with it the glamour of "as seen on TV." Programs did not cost very much, and they could be turned out overnight in-house.

Most programs were not produced for duplication and distribution and they usually were played back on the same equipment they were recorded on. Interchangeability was still a problem and playback was generally confined to the training area where the tapes were tied into classroom lectures or self-paced instruction using workbooks and other audiovisual media.

By now you may have noticed that things keep coming full circle and history has a way of repeating itself. We are now seeing the same thing happening all over again as video users in field locations make their own tapes for local use with inexpensive consumer video equipment.

Setting the Stage 17

An obvious success in the training area, video was quickly sought after for other applications within the organization, often Sales Training because the medium communicates people better than it does anything else.

The Awakening Giant

Program distribution was always a problem until the standardized 3/4-inch videocassette came along in early 1972 and private televison networks were born. The slumbering giant awakened.

In its early years the U-Matic videocassette was used primarily as a means of program distribution, even though most machines came equipped with a recording capability. At that time the programs distributed on videocassette were generally produced on another format such as film, two-inch or one-inch videotape, or, even, 1/2-inch, open-reel videotape.

A steadily improving technology soon brought about low-cost cameras and 3/4-inch cassette editing systems. Many organizations which began video solely with distribution networks started buying this equipment and producing their own programs in-house. A new generation of "user/producers" was born, again attracted by the speed, convenience and economies of the videotape medium.

This development was aided considerably by the rapid adoption of the 3/4-inch videocassette format for broadcast news gathering (ENG) as a replacement for film. Just as rapidly, corporate users adapted the ENG techniques to their own productions which now no longer required expensive fixed studios.

Beginning as a distribution format, the 3/4-inch videocassette evolved rapidly in the second-half of the Seventies into a production workhorse and it is still going strong, as the survey results in Chapter 9 indicate.

Into the Eighties

But, paraphrasing the old saw, "If you don't like the present format, wait a few minutes."

In the early Eighties 1/2-inch videotape returned once again, this time in cassette form and in not one, but two, formats: Beta and VHS. This time, however, it found its niche in the consumer market. After 15 years and a dozen different products and formats home video was born at last.

Along with it, the new half-inch formats brought a horde of inexpensive cameras and VTRs, both into the home and into the workplace. Now video is everywhere, as the 1985 year-end cover story of *Newsweek* trumpted. Inexpensive video equipment can now be found just as readily on the job as at Baby's first birthday party.

The predictions made by industry expert David Lachenbruch in our 1977 study have come true:

"With videocassettes — eventually video cameras — becoming commonplace home appliances, the "gee-whiz," wonder-of-the-future aspect of video will quickly be dispelled. When officers and board members of corporations and school and institutional officials have video recorders in their own homes they will see the value of the medium far more clearly. Private television then is no longer an electronic wonder or an elaborate substitute for a slide projector, but a valuable and accepted tool that does an unparalleled job in communications.

"It seems inevitable that consumer-type hardware and formats will spread to many video communications networks, simply because they are so economical.

"Eventually, consumer tape players will make possible home-office, home-plant and home-school interfaces, permitting the viewer to take some of his audiovisual work home with him — a definite advantage to both the producer and the user of video material."

Chapter 2
Stone Tablets to Storyboards

All Communications Is Essentially Visual

Everybody these days seems to be talking about the revolution in communications and the impact of the new communications technologies. Modern Paul Reveres gallop through the land yelling: "The digits are coming! The digits are coming!"

While the rapid change in technology is not enough, the deregulation of the telecommunications industry has left everyone in a tizzy, making it virtually impossible for most people to keep up with all that is going on.

Now is the time for everyone involved in communications to stop frantically chasing the latest technological developments and reassess what communications is really all about.

In the simplest of terms: *communications is the process of getting a thought or idea out of one head and into another.* This is the most difficult thing any human being has to do, and he or she is faced with this challenge almost every waking moment from birth to death. To do it, we have developed a complex array of techniques and systems, often called *media*. Many of these, unfortunately, have become ends in themselves.

Before we proceed any further, let's go back to the basics and then try to make sense out of all of the technology. A recently released book featuring the thoughts of prominent CEO's on the future of business really hits the mark. Mitchel Ford, former Chairman of Emhart Corporation, implores:

"We need to back away from our traditional preoccupation with the trappings of communications — the hardware, which has taken on a deity of its own — and take a fresh, objective look at <u>what</u> and <u>why</u> we are communicating."

To begin with, all thought is essentially visual. We think primarily in visual concepts. No one really thinks in words, numbers, bits, digits, sounds, smells and so on . . . at least, no one you would want to have over for dinner.

Therefore, all communications begins in a visual form and ends, if all goes well, in a close approximation of that same form. The challenge of communicating is to get one's own inner visual image or concept into another person's mind with a minimum of distortion.

In our everyday speech we frequently make reference to our mental images. "I *see* what you mean," for example. Successful leaders are described as people of "great vision". We are told by personal achievement motivators to "create a clear picture of our goals and objectives" and to "keep the image in front of us" as we pursue them.

We have become extremely adept at using our linear words to convey our mental images. In setting the stage for Shakespeare's epic "Henry V", the Chorus tells us to . . .

"*Think* when we *speak* of horses that you *see* them, printing their proud hooves i' the receiving earth."

A travel writer communicates the feeling of a distant countryside with:

"Ducklings the color of lemon drops dither down Main Street, followed by an unruly blur of cats, geese and goats. The wind is still stiff, and blows the crows around like cinders over the dry fields."

Since we have never been too successful in transmitting detailed images in living color directly from one mind to another, this is where media comes in. The purpose of a communications medium is to break the visual image in our minds down into some system of components that can be delivered to another person (or group of people) for reassembly into a replica of the original image.

We do this in face-to-face communications as our primitive ancestors did, with sounds, signs and facial expressions. To this day, this

Stone Tablets to Storyboards 21

informal media system is still the most preferred means of communicating. *Formal* media systems had to be developed where face-to-face communications were not possible because the receiver was displaced from the sender in either distance or time.

Much has been written recently about the differing functions of the left and right hemispheres of the human brain. The right hemisphere which deals in images and abstract concepts is essentially non-linear in its processing. The left hemisphere, which holds the speech and reading centers, seems to be linear in its processing and does the translation work needed for us to tell someone else what our visual images are all about.

Incidentally, this divided function was the basis of a recent thriller novel wherein the protagonist has secret information stored in his right hemisphere which has been surgically separated from the left. Theoretically, he has no way to get it out. Meanwhile the bad guys are after him trying to access his right brain. He resorts to some interesting visual communications techniques to try to get the stored information from the right to the left side so he himself can understand what it is that's locked in his head. An interesting study in internal communications, to say the least!

This split-brain phenomenon in humans seems to be one of the stumbling blocks in the development of so-called artificial intelligence in computers. Computers are linear by nature. They can't manipulate non-linear abstract concepts as a human can. They can't think visually. Until this problem is solved, man will still retain the upper hand.

Jean Auel in her wonderful best-selling novel "The Clan of the Cave Bear", portrays a society of Neanderthal cave dwellers who convey rich visual images swiftly and graphically through a sophisticated sign language. Physically they are incapable of articulate speech. They use sound in communicating only for emphasis and to gain attention.

Then along came the Cro-Magnons with their capability for speech and the nature of face-to-face communications changed radically. Words became the primary medium and visual communications became secondary. However, the information flow was greatly expanded since man could now communicate on two channels simultaneously. In effect, as a video engineer of today would say, the information bandwidth was greatly increased.

Enter Media Technology

Interestingly, it was the Cro-Magnons who were responsible for the magnificent cave paintings at Altamira, Spain and other locations in southwestern Europe some 25,000 years ago. Obviously, the possession of verbal skills in no way inhibited their visual imagery.

These cave paintings may be the first formal media systems we have any record of and predate writing by some 5,000 years. The prevailing theory is that these paintings were a form of religious magic, communications between man and someone, or some *thing*, he couldn't see — a god or a spirit. By painting a picture of a successful hunt on the wall of a cave for that unknown essence to see, the caveman was, in a sense, placing an order for the next day's lunch.

Commenting on the paintings, Flora Lewis of the New York Times said in a recent article: "Perhaps drawing them was a kind of incantation, a belief that the capacity to record an image would promise success in the most difficult of hunts. The power to portray can be seen as the power to dominate."

Of course we could raise the philosophical question at this point as to whether or not our corporate video programs are merely a hi-tech version of the same sort of religious magic. Do we create programs out of the blind faith that someone, or some *thing*, will view them and carry out the program's communications or training objectives?

These magnificent cave paintings come as close as anything man has done since to duplicating in another's mind an individual's own mental concepts. As a media system, however, the paintings have two serious drawbacks — it's tough to make copies of them and they are impossible to distribute in their existing format.

For several thousand years, communications media were restricted to some form of pictorial representation, first in magic and religion and later, as society evolved, as a means of communications and record keeping among men.

One of the earliest forms of writing was used in ancient Sumaria, later Babylon, some 5,000 years ago. At that time, four simple lines scratched on a soft clay tablet by a sharp reed stylus indicated an ox. A far cry from the fully detailed cave painting, it nevertheless effectively communicated a whole visual concept, and, as our video engineer of today would say, required much less bandwidth. Everyone who saw the symbol knew what it meant. It was impossible to be illiterate when

Stone Tablets to Storyboards 23

using simple pictures.

But the pictures did not stay simple. As the use of the clay table/reed stylus medium expanded, the messages became more complex and the symbols became more stylized. The medium itself demanded certain changes and altered the way the messages were presented.

The medium did not become the message, but it had a major impact on how the messages were presented. The pointed stylus that left a groove and a raised ridge was discarded in favor of a stylus with a triangular tip that was pressed into the clay. Vertical columns on clay tablets were not "user friendly" to the scribes, so they changed the whole system to horizontal rows for speed and legibility and to avoid smudging. Technological obsolesence was invented and "symbol-processing" entered the "stylus pool."

With these innovations, the pictographs changed from simple drawings of objects to abstract symbols that were easier for the scribes to write in a hurry. You had to be able to read to understand them. Furthermore, there was no direct relationship between the written symbol and audible speech. The association between the sound of a word and its symbol had to be learned. Thus, *illiteracy* was born!

The audio and visual portions of the message were not linked together until the Phoenicians invented the alphabet. From that time on we became very literal and, as Marshal McLuhan claimed, very linear in our communications and in our thinking.

Not really. The alphabet merely became a means of transmitting thoughts and ideas faster and more efficiently. William Shakespeare and thousands of other writers have proved that our best writing is that which evokes the greatest mental images. This is something that is usually overlooked in most corporate literature.

While we are primarily visually oriented in our thinking and in our communications, historically it has been difficult, time-consuming and expensive to record and transmit ideas in picture form. It is much easier to convert images into words, symbols or even digits which can be turned out and transmitted in vast quantity.

Most forms of communications involving media consist of someone taking a visual idea and converting it into non-visual form — such as words — for transmission and storage. The receiver then must convert the non-visual elements back into visual concepts. A newspaper report of a fire or some other dramatic event is the most familiar example.

The fact that we can go through this process easily and at incredibly high speeds should not obscure the fact that more often than not we start and finish the procedure in *visual* concepts rather than abstract verbal concepts.

The Picture Changes

Starting with still photography, later motion pictures, and now television, the picture has changed drastically. (No pun intended.) First, we had photojournalism — the reporter with the camera instead of the pen who was the backbone of the mass circulation picture magazines, *Life* and *Look*, until another more convenient picture medium, television, knocked them off.

Ironically, for television's first 25 years we usually watched the written word being read to us. Then photojournalism became fully electronic and the video camera now plugs the viewer directly into events as they happen. Goodbye reporter and news reader. The video camera now "writes" the story.

This illustrates a trend that is, in fact, revolutionizing all human communications. Since the cave paintings over 10,000 years ago, *all use of communications media has been at the convenience of the originator or sender* — rarely in the form most convenient to the recipient. As we said, it was expensive, time-consuming and difficult to communicate in picture form in the past. Given a choice, however, most people would rather get their information visually than verbally.

More people get their news and information about what is going on in the world from TV than from all of the news magazines and newspapers combined. In the latest Roper Organization study for the Television Information Office (TIO), respondents maintain — for the 24th year straight — that television is their most *believeable* source of all news as well. Television's credibility as the prime news source now wins by a two-to-one margin.

Aside from whether it is right or wrong, the fact is people support and use the medium that is most convenient for them. In the Roper poll they even agreed that watching commmercials on television was "a fair price to pay" for being able to watch the programs. They pay this price seven-and-half hours a day on the average according to another study.

With the avalanche of new technologies, the most important thing for anyone in any area of communications to consider is that the *control of the communications process is shifting rapidly from sender to*

Stone Tablets to Storyboards 25

receiver. *That is, many of the new technologies enable the receiver — rather than the sender — to initiate the communications process!*

For the first time since the cave paintings *we now have the means to communicate with anyone, anywhere, at anytime and on a cost-effective basis.* With the process being initiated by the receiver, the only problem facing the communicator is content. And it had better be something that produces a mental image quickly and easily.

Adrift in the Data Stream

All of this leads us to the question of what effect the new technologies will have on basic human communications and our need to communicate our visual images.

The world is rapidly going digital . . . the language of the computer. Increasingly we are using bits and bytes to convey our thoughts. The last working linotype machine was mothballed recently as the entire publishing world rushes towards automation. Words are now swept along in data streams by machines that cannot distinguish them from bank balances or telephone charges.

Words-on-paper as a primary means of transmitting and storing information is starting to be phased out of our society. Libraries are seriously considering becoming electronic information data bases. Newspapers soon may never know the bottom of a bird cage or be used to wrap fish if Knight-Ridder, Gannett, Dow Jones and other publishers' experiments in electronic distribution bear fruit.

Many companies are looking at ways of moving the contents of "overnight letter" packages without moving the packages themselves. Dial-up, common-carrier computer networks with built-in protocol conversion are spreading like giant spiderwebs across the land, linking Apples to PCs to Commodores on thousands of desk tops all over this world.

As we have seen by the evolution of graphic communications from Altamira cave drawings through clay tablets, hieroglyphics and other ideographs to the alphabet and movable type, the technology of a medium influences the *form* of the message as well as its reception by the received.

Can the data stream communicate the rich visual imagery that words have conveyed over the centuries? James Joyce and Henry James are best suited to the printed page. Difficult in print, it would be sheer torture to try to get through the first sentence of "Ulysses" on a CRT screen. The format just does not lend itself to that kind of expression.

Does this leave video as a shimmering visual oasis in an otherwise barren Sahara of data and digits? Yes and no. For some time we have been fond of saying: "Video is dead!" at ITVA chapter meetings around the country and at other gatherings. This produces some sweaty palms among video users and producers, especially those who have just put a lot of money into equipment. What we mean, of course, is that video *as we know it* is dead.

Video really isn't dead . . . *it's data*! Video is following words into the data stream and it is going to be manipulated and shaped by digital technology like everything else.

Pumping Video

During a video needs analysis study for a major electronics industry manufacturer we ran into a data systems manager at a plant deep in the tobacco fields of North Carolina. In addition to having what is probably the most advanced computer and internal data transmission system you will find anywhere, the data manager has a complete ENG video system. He says he uses it to "capture data".

"What are you talking about?", we asked him. "What do you do, go around videotaping bits and bytes on computer terminals and printouts?"

"Naw," he drawled. "We take pictures . . . you know, *data*. A video picture isn't anything but a bunch of data. We just capture it and pump it around the place like we do the computer data. We have to move big batches of data from machine to machine, so video isn't any big deal. Once you have the system in place for high speed data movement, video's a cinch."

This data systems manager is a rare bird who is equally savvy in data processing, telecommunications and video. There should be more like him . . . "a man of *vision*", as it were.

Within the next few years, he intends to have high resolution color video displays at every work station and computer terminal throughout the plant. This is in addition to the TV sets now dotting the shop floor which are used to monitor and control the computer-driven machine tools that turn out high-precision steam turbine blades. Anyone at any desk or work station will be able to receive a full-color, full-motion television picture in addition to a full range of computer displays.

With this kind of capability the mind runs wild with possibilities. Important company messages seen by everyone throughout the plant at the same time with no one having to leave their work stations . . . dial-up access to libraries of training tapes . . . videoconferencing . . . the potentials are endless.

If video becomes data, who will be responsible for it? Right now not enough corporate video people are directly involved in the new communications technologies that are changing the work place. What's happening in video teleconferencing is a prime example.

Our own first involvement in what we now call "Private Television" was with big- screen, multi-city videoconferencing in 1969. There were few videotape networks then. Most video productions that were to be seen in many different locations were produced live and distributed by telephone lines at great expense. Professional video people were involved from end-to-end.

With the advent of the videocassette, videoconferencing went into a decline. Today, it is a hot topic but it has taken a new twist. Data technology now makes it possible for the video signal to be compressed from 90 megabits per second (Mbps) to 1.5 Mbps, and thus be carried on a standard T1 data communications channel like any other piece of computer-generated information.

Company videoconferencing rooms are springing up all over the place, frequently designed, installed and operated by the Systems and Telecommunications people. In some organizations, the video people do not even know it is going on.

What our South Carolina data systems manager is operating is a "local area network" (LAN). This is really nothing more than an in-house coaxial cable system capable of carrying hundreds of times more information than ordinary twisted-pair telephone wires. LANs are the core of most office automation systems. Some video people don't know anything about them. They should. Chapter 15 covers this in more detail.

Video as we've known it and practiced it in the past may soon become the electronic version of the Altamira cave paintings — beautiful, expensive, impractical — and a thing of the past.

Our mental images are going to have to be carried by the new technologies just as the image of the bull on the wall at Altamira was later communicated by three grooved lines on a soft clay tablet. However, we still must have the creative thinker — "the person of vision" — to originate the concept that must be recreated in the mind of the receiver. He or she merely has to learn a new technology to do it.

This is the future role of today's video person.

Chapter 3
How Organizations Communicate

When a new organization is started, communications is rarely a problem. After all, everyone usually knows everyone else and the lines of communications are very short and direct. Most information is conveyed through face-to-face contact or via the telephone. There is no time or need for lengthy memos, newsletters, bulletin boards, videotapes or other formalized media systems.

During the heady period of start-up, everyone takes a solemn oath that they will do their very best to keep everyone else informed and that things will not fall between chairs.

However as the organization grows and becomes successful, communications invariably get worse. Direct personal lines of communication get stretched and broken as expansion disperses people and leads to one reorganization after another. Everyone assumes that important information is being passed along through the chain-of-command. Unfortunately in too many cases it is not.

No one is really to blame. It is a simple fact of organizational life that must be recognized. At some point along the way, hopefully before things get too far out of hand, something will have to be done about it.

What usually happens is that some bright up-and-comer gets overly enthusiastic about some new communications technology and tries to ram it through the rest of the organization as the panacea for all that's wrong. Sometimes it works and things do get better. But too often it fails due to the lack of proper planning and the situation ends up worse than before.

In the course of our work we have interviewed hundreds of executives at all levels from board chairmen to first-line supervisors in dozens of organizations in all kinds of industries. To a man, and woman, they have all said that *communications* was one of the biggest problems facing their organizations.

Each expressed an eagerness to help do something about it . . . if only "they" were willing to recognize the problem and do something first. Most felt a major problem was that there were few viable systems in place within the organization to enable communications to occur quickly and efficiently.

One board chairman summed up his own situation quite succinctly:

"When it comes to operational problems we will devote hours of management time in meetings in an effort to solve them because we know they can be solved. Communications is one of those subjects that we all realize is important but we never spend more than 10 minutes discussing because nobody feels it *can* be solved."

There is always finger pointing among lower level managers toward upper management where the lack of information is perceived as top management not caring. Lower levels tend to regard upper management as being "insensitive" and unwilling to communicate. "They don't tell us, which means they don't care!" Many middle managers and first line-supervisors feel they are not getting enough information to be effective as communicators.

Upper management seems willing to communicate but does not know how or often lacks the proper means to do so effectively. Middle management points the finger at "them" but fails to regard communications as an important part of their own jobs. To the hourly worker, the immediate supervisor is "management." There are strong feelings that these managers are poor communicators and the workers assume that the top is as well.

At all levels there is a strong desire for faster and more candid means of communications. Coupled with the recognition at upper levels of the need to *send* information is the strong need at all levels to *receive* more information about the company and its activities. The need to know the reasons behind company policies and decisions was voiced consistently throughout the interviews. It is generally agreed that in most cases there was no clear policy on communications coming from the top.

The Rumor Mill

Unfortunately, for most employees the primary source of fast information about their organization is the rumor mill. As everyone has experienced, the rumor mill is extremely fast and nearly always beats company announcements. Even though its accuracy rating is really only something like 60 percent, it always seems to be accurate in reporting *actual* events and therefore it is believed *all of the time*.

This is a serious problem because the speed and accuracy of the rumor mill weakens the credibility of official company communications. Rumors have the advantage of being transmitted by direct personal, face-to-face communications, always the preferred way of getting information.

Company information too often is transmitted only in print and is usually too late in coming. As one somewhat embittered plant worker told us: "If you want to know what's going on, you go ask 'Rumor Control'. The company don't tell you nuthin."

The rumor mill is particularly galling to managers in the field who are caught off guard by employees bringing them important company information that they heard outside the organization before it has been disclosed internally. Traditionally most employees want to get company information through their immediate supervisors. When the supervisors are thought to be uninformed, their authority is undermined.

Credibility

The largest single communications challenge facing most organizations today, both internally and externally, is one of *credibility*. "We hear the words, but we don't believe it!" sums up the feelings of most non-management employees who seem to share the feelings of an organization's publics.

Credibility is based on a number of interrelated factors that influence the person receiving information. Among others these include:

- the source;

- how willing the receiver is to accept the information;

- how far the receiver is removed from the source;

- how the information is delivered.

Being honest and telling the truth is no guarantee of credibility. Trust is involved and trust is based on *consistency*. That is, the source must have a perceived reputation with the receiver for being consistently truthful and honest. This, of course, implies more than a one-shot effort at truthful communications. There must be continuity.

Communications never take place until the information is accepted by the receiver. When a person wants to believe what he is being told, the source automatically becomes credible. Accepting information does not always imply agreement, however. We often disagree with people we know to be honest and sincere. Our disagreement does not necessarily challenge their credibility.

At best most organizations are regarded as being honest but their messages are not being believed. As one local manager told us recently: "My friends are still my friends and they know I work for the company. None of them agree with me on the controversial issues facing the company, but they are at least willing to listen to me. That is the best we can hope for at the moment. We can't change people's opinions until we can at least get them to listen. This is our priority communications challenge."

Distance from the source can mean several things. Communications direct from the source to the receiver is always more credible than information that is relayed second and third hand. Printed forms of communication always put distance between the source and the receiver. Then there is the distance of backgrounds and position that can also occur even in direct face-to-face communications.

Media systems or communications channels have their own built-in characteristics. Again, face-to-face communications generally are perceived to be the most credible. Most people like to think they can tell when someone is telling them the truth or not.

Recent surveys have again confirmed the fact that people are more apt to believe news they see on television than what they read in newspapers. Pictures, still or moving, are *direct* information. Printed information goes through many hands and processes from source to receiver and is not perceived as being direct.

Perceptions of Video

We have found that in most organizations that were not using video extensively for communications there is a strong desire from the top on down to use it as a means of solving their communications problems.

How Organizations Communicate

This was particularly the case in those organizations already using video for training and where people were satisfied with its effectiveness.

Unfortunately there is the risk that video may be regarded as a remedy for all communications problems. This enthusiasm for video sometimes stems as much from a lack of familiarity with all that is involved in the medium as it does from the hope that somehow it will make everything better.

Certainly video does solve a large number of communications problems. Next to face-to-face contact, it is the most direct form of communications. It also is far more credible than print media. To many top managers it is a more *efficient* form of communications.

One company president told us:

"Overall, our communications have improved greatly in the last several years. The problem is that things are moving much faster than ever before. There is so much more that needs to be gotten out to the employees that we're now further behind. *What we need to develop are communications channels that by-pass the various levels of management and get information directly into the hands of our employees. As I see it, video is an ideal way to do this.*"

Other top managers like it because everyone in the organization gets the same message. Policies and directives are no longer subject to filtering, distortion, embellishment or blocking at various levels of the organization.

Hourly employees like video for the same reason. They get their information straight from the source. Because a video "memo" can be gotten out, literally in a matter of hours, they no longer have to rely on the rumor mill for their information.

Most important, however, is the fact that in most cases the audiences appreciate the *effort* of video communications as much as they do the content of the programs themselves. "It shows they really care," one plant worker told us.

Chapter 4
Determining Communications Needs

Video Is Not A Fad

A great deal has been written in the last three years about management styles or theories which have often become "fads."

In some companies managers have lived through *zero-based budgeting*, *intrapreneurship*, *management-by-walking-around*, *one-minute managing*, and *Theories "Z"* and *"Y"* in rapid succession. One field communication manager in St. Louis said that by the time he had mastered the techniques of one management fad, it had been superceded by yet another. He said he felt top management did nothing other than determine "the management theory of the week" with little thought about follow through or the impact of a particular fad on the entire organization.

When video first came into serious use in many organizations, field managers thought that it, too, was just another "fad" of top management's and that it would soon fade from sight.

Certainly by the longevity of most video units as seen in Table 6-2, video has outlived most of the management "fads" that have come along in the last twenty years. One thing that has kept video alive and well and a sure "fit" in any corporate culture is the fact that video from the start was seen as a solution to a definite corporate problem and was supported by top management at its introduction.

As we and others have said repeatedly over the years, unless there is a sincere commitment from the top and a well thought out plan for its use, even video could become a "fad." When perceived as a "fad," no one takes the medium seriously and it becomes not only a waste of time and money but of the confidence of the organization's employees or members.

In organizations which have been using video successfully over a long period of time, you will find that it is a *planned* activity. All media and messages for communications, training and marketing are organized on a fully integrated "systems basis." Publications, newsletters, bulletin boards, A/V presentations, audiocassettes even purchased advertising and sales promotion materials prepared for external use, all work together to achieve clearly stated, previously determined objectives. Media usage is not a scattershot activity with little coordination between different people handling each function. Even the *informal* or non-institutionalized forms of communications are better developed in these organizations.

Management on all levels take their job as communicators more seriously when presented with concise communications' objectives and a carefully conceived plan for implementation.

Communications Needs Analysis

The video medium is highly cost effective when its use within the overall media system is explicitly analyzed. It is simply too expensive and demanding to be used casually and without adequate preparation.

It is absolutely essential for any organization planning to use video to do a full-scale, organization-wide information and communications needs analysis. Even organizations which have been using video for an extended period need to reassess their needs periodically. Times change and so do communications needs. It is never too late.

Nevertheless, it is disappointing to find that only 46 percent of the current survey respondents have done a communications or training needs analysis study.

While we are using video as the reason for a needs analysis, such a study also will help evaluate the media currently in use. It also is just as valid for the introduction of any new media technology in the workplace.

In essence, a communications needs analysis should do the following:

1. Evaluate existing formal and informal communications channels and media systems and assess their effectiveness;

Determining Communications Needs 37

   2. Analyze the message flow and content;

   3. Assess the overall communications "climate" with the organization;

   4. Define and profile the various audience groups within the organization and determine where they are located; and,

   5. Analyze the organization, staffing, policies and procedures of the communications operation within the organization.

The result should be a series of recommendations to management which cover . . .

- Communications objectives and policies;
- Information subjects and priorities;
- Media systems, such as print, bulletin boards, video and so on;
- Allocation of which messages to which media;
- Media production and distribution systems;
- Organization and staffing necessary to do the job;
- Systems for budgets, cost accounting and cost analysis;
- Systems for feedback and program evaluation; and,
- Use of in-house capabilities vs. outside services.

Communications Task Force

A communications needs analysis study works most effectively if it is carried out through an in-house study group or task force of senior officials. This team should be commissioned by top management to look either at the entire corporate communications picture or merely at whether video or any other new technology should be used by the organization. This task force concept can be used to evaluate the impact of the communications department and/or video technology as well.

It is always more effective if the actual study itself is carried out by an organization working directly for the task force. Unless the task force members are severed from their regular jobs for the duration of the project most will not have the time, not to mention the expertise, to do the job properly.

The rationale behind establishing a corporate-wide communications task force is to provide the needed input from an organization's wide variety of staff and line constituencies. Also, it defuses from the outset possible political conflicts that can arise over communications policies and procedures. The study and any ensuing recommendations should not be identified as being the product or program of any one division or department within the organization. That is the fastest way to get it shot full of holes.

In addition, it provides a front-end involvement on the part of many who may have a say later in the approval or implementation of any new system. It also speeds adoption of the proposals by top management.

Information Channels

We have identified some 20 information channels — both formal and informal — within most organizations. On the informal side these include:

> Face-to-face (one-on-one)
> Telephone
> Meetings (scheduled)
> Meetings (unscheduled)
> Travel
> Letters
> Memos
> Telex
> Reports
> Carbons of any of the above

The formal, or institutionalized, information channels include:

> Company periodicals
> Company booklets
> Bulletin boards
> Audiovisual presentations
> Video programs
> Electronic databases
> Industry publications
> General business publications
> Newspapers
> Radio and broadcast TV

Next, it is important to identify and categorize the various kinds of information that flow through these channels. These include . . .

> General business information
> Industry information
> General company information
> Corporate policy information

Determining Communications Needs

>Systems and procedures
>Day-to-day operating information
>Public affairs information
>Marketing, advertising and sales information
>General information on individual divisions and departments within the organization
>Personnel policies
>Benefits
>Employee activities
>Personal development
>Job and skill training
>. . . to name but a few.

Nearly all originate at different points in the organization, require greater or lesser preparation effort and have shelf-lives ranging from 24 hours to two years or more.

Information Categories

Most of the above, however, fall into one of these broad categories which are common throughout most organizations.

1. *Organizational Policy Information* — This is the all-important information on how an organization conducts its affairs in various areas. Included are organizational goals and objectives as well as the strategy involved in achieving them. Policy information necessarily emanates from top management and is normally disseminated throughout the organization via the "chain-of-command." This is generally *need-to-know* information.

    Dissemination of policy information in most organizations is largely by directive, memos, executive presentations and meetings. These are frequently supported by in-house publications, audiovisual presentations and/or video programs.

2. *General Organization Information* — Not critical to the performance of any manager's or employee's job is a wide range of nevertheless highly important information which provides vital background on the organization's activities in general (e.g., sales and earnings, expansion plans, etc.) and various operating and staff components in particular (e.g., customer services, transportation, personnel appointments, reorganizations, marketing and merchandising plans, etc.)

    Often this is described as *nice-to-know* information as compared with *need-to-know* information. Because of its discretionary nature, this is the information category that is most often neglected by management. It requires a clearly stated policy and commitment from the top in order for it to be gotten out. It is usually disseminated via formal in-house media systems, that is, publications, video, bulletin boards, etc. as opposed to distribution via memos, reports, meetings, presentations and so on.

3. *Operating Information* — This is the day-to-day flow of information within and between components that is directly job-related and ranges from operations and production to sales and marketing information. It also would include specific operating and implementation procedures related to policy decisions. Origination is usually within line operating components.

Dissemination is usually very direct, for example, face-to-face, direct correspondence, forms or telephone. Formal media systems are seldom used for this kind of information.

4. *Administrative Information* — This consists of all information related to administrative and personnel activities — the so-called "housekeeping" or "bread and butter" functions — such as salary administration, performance evaluation, benefits, job and skill training, sales training, management and human resource development, safety, regulatory compliance, employee services and so on.

In most organizations this information originates in a variety of specialized staff support areas and is disseminated initially by memos and directives. However, successful implementation usually requires the use of one or more in-house media system as well.

5. *General Business, Industry and Public Affairs Information* — This information generally originates outside the organization but is disseminated internally to specialized and/or general audiences. Subject matter ranges from economic information (and education) to environmental, legislative and consumerism issues. Dissemination is through a variety of institutional media systems but is sometimes conveyed directly to employees by top management when the subject is important enough.

Information Shelf-Life

The timeliness, urgency or "shelf-life" of the information varies by category as well as within categories. General organization, industry, or business information is usually disseminated within 24 to 48 hours and has a shelf-life or about one to two weeks.

Policy information is generally disseminated with 24 hours, sometimes within hours if it is of a critical nature. Its shelf-life is indefinite and it will last until either cancelled or superceded. Transmission of operating information is either instantaneous or within 24 hours in most cases.

Information on procedures, products and services varies in speed of distribution and can require either immediate action or be stored indefinitely (e.g., informational materials on products and services).

Administrative information, such as an explanation of company benefits, takes the longest to prepare but usually has the longest shelf-life. Dissemination also may take longer than other information categories.

Where the preparation and production of material for in-house or institutionalized media systems is not involved, the originator usually prepares the material that goes out. Where media is involved, in-house communications professionals generally prepare the materials in consultation with the originator or "content specialist" in most organizations.

Information Flow

The most important thing the communications task force should do is get out into the field and talk to the people receiving the information, at all levels. This includes talking to the chief executive as well. More than one CEO has complained to us that he feels cut out of the information flow within the organization and that he has no idea whether the word is getting out or not.

This reflects the most common finding of any needs analysis study — the uneven and inconsistent flow of information within an organization. *Not everyone gets the word and not everyone gets the same word.*

In organizations which rely on the word-of-mouth, pass-along system of communications this is the biggest problem no matter how well-intentioned or open in its communications management may be. It invariably results in pockets of isolation and feelings of "being left out". This, of course, plays hob with morale . . . and, consequently, productivity.

The best intentioned communicators in most organizations are the chief executive and the first-line supervisors. Between these two is a communications wasteland called "middle management."

Lower middle management, from department heads on down, are the worst communicators in any organization. In a sense they are the "forgotten people." They receive little or no information on corporate policies and direction, yet, ironically, they are most often the ones charged with the interpretation or implementation of it.

Few middle managers see communications as a part of their jobs. Most are preoccupied with playing politics or slavishly observing the rules of "the system". In a period when the ranks of middle management are shrinking, most of their activities are directed at preserving their jobs and pleasing their bosses. Little thought is given to the actual goals of the organization that employs them and possibly for a very good reason — they are not privy to what those goals, objectives or commitments are.

Withholding or releasing information is more often considered a form of power not a means of getting the job done. We have called this group the "Berlin Wall of Communications" because it is so difficult to penetrate either upward or downward.

Yet time and time again theorists and pragmatists alike admonish that "communications in times of rapid change become critical." And rather than narrowing the gap between top and middle management it is actually becoming wider.

Case in point: a well-managed company we know prepared a simple, clear-cut set of corporate goals and objectives that were to be distributed throughout the organization. The Management Policy Committee communicated the goals to the Management Planning Committee — made up of all operating and staff vice presidents — who in turn passed the information along to the next level of management.

Later, in discussions with first-line supervisors, the key links in getting the word to the rest of the troops, we found that none had ever heard of the new goals and objectives. Tracing the breakdown in communications we found that virtually none of the department managers communicated the information to their staffs because: "no one told us we were supposed to" . . . "we thought it was company-confidential." Needless to say, the chairman was appalled when we reported this to him.

Determining Communications Needs 43

Most senior managers realize the communications problems inherent in the layers of middle management. That is why the effective use of media systems provide them with a means of tunneling through the obstruction . . . or going around it.

Good managers are instinctive media beasts. They are attracted to and will use media systems that enable them to communicate — and therefore manage — more effectively.

This is one of the reasons for the successful growth of private television. Top management has seen it as a consistent means of getting the right information distributed throughout their organization. Policy annunciated at the top gets the same understanding all the way down throughout the organization. No longer can a self-serving middle manager bend interpretation or constrict knowledge to his or her own ends.

Chapter 5
Management and The Tube

The level at which the decision was made to use video remains about the same as it has been for the last 12 years. The decision was made at the very top. Again more than 50 percent of the sample reported that the CEO, President or Chief Administrator had made the ultimate decision to use video. (Table 5-1)

TABLE 5-1

Management Level At Which Decision Was Made To Use Private Television
N = 240

Mgt. Level Making Decision	% of Total
Chief Executive Officer, Board Chairman or President	51.7%
Exec./Senior Vice President	21.2
Vice President/Division Head	20.0
Department Head	6.7
Section Head	0.4
	100.0%

There is a very significant drop below the decision-making level of Vice President/Division Manager, however. Ninety-three percent said the decision was made at that level or higher. In 1981, only 86 percent answered similarly. Lower level managers — department heads and section heads — have lost some of their decision-making powers over the medium.

At the same time, the significance of the top management's support and influence over video is seen in Table 5-2, which shows that 91 percent of the heads of organizations appear on video either sometimes or often. In 1981 that total was 84 percent.

TABLE 5-2
Appearance By Head of Organization on Private Television
N = 246

Appears	% of Total
Never	8.9%
Sometimes	66.7
Often	24.4
	100.0%

The people-oriented non-industrial segment finds more CEO's on television with 94 percent making appearances, while only 88 percent of the industrial CEO's use the medium. In the medical, non-profit and government sectors, 82 percent of the heads of organizations make frequent appearances on private television.

Albeit some of the CEO's apperances may be cameos or guest shots, where he or she may lend an appearance to boost the importance of the message being delivered, we find that top management is the third-ranked source of video programming. In other words, top management is now originating a great deal of programming as well as just making appearances.

Proposals to Management

No matter how many years of experience one has, making a proposal to management is an awesome task. To tackle the problem and feel that you have at least a chance of being heard and your proposal considered, you must approach it as you would any other communications problem you have to deal with. That is, you must find answers for the age-old question: _Who do you want to do what and why . . . and how are you going to get them to do it?_ In other words you must analyze your audience, which in this case is management, set clear objectives and determine what the message and medium should be in reaching this audience.

Do your homework. What are your organization's goals? Outline how video can aid in meeting these objectives by an increased use of the medium, better equipment, more playback units, additional staff, a larger budget, etc. At least you will show management that you have their best interests at heart.

Find out what other organizations are doing. The following chapters of this book are designed to help you in this area. Also, attend ITVA meetings in your area to share experiences with other video users.

To continue your audience analysis, these questions need to be answered. What is the prevailing attitude toward communications in your organization? Toward video? Is communications a pawn in an internal power struggle or is it in a neutral corner? What axes will be ground for or against your proposal because of this?

Can you go directly to the top decision makers? Or, do you have to go through a number of intermediate layers? Are you playing to someone's ego or the balance sheet? Either will play an important role in the acceptance of your proposal.

What is the current corporate culture? What "language" is being spoken in your organization? Find out and speak *management's language.* (While on the subject of language, do not use equipment model numbers or technical jargon in any management proposal! Top management has no interest in the technical details.)

What form should your proposal take? Top executives are impatient with lengthy documents. Middle management wants to see all the details and all of the documentation. If this is the case in your organization, prepare two documents: 1) a short "Executive Summary" of what you are proposing and why, what it will accomplish and how much it will cost; and, 2) a separate, detailed plan with all of the supporting materials.

Your recommendations should be oriented towards solutions to problems, not just hardware and facilities. If possible, use examples. If you are just starting out in video, you might find a friendly manager who would like to do an experimental pilot program (which he would pay for) that you could show to the rest of your organization. You also might collect some sample videotapes of a similar kind done by other organizations. ITVA members, your video dealer or a local production house may be of help here.

If your plan includes equipment or staff additions, be sure you provide a rationale for the expenditures. For example, you might want to outline the number of additional programs you will be able to produce for particular clients who have been on your case. Or, you may need to explain the benefits to the organization of having more locations able to show Employee Orientation or Benefits programs on demand rather than sending for the tapes and then renting or borrowing the playback equipment.

If its staff you need, you had better have job functions, career pathing and salary administration figures in your proposal. Further, you may need to do some cost comparisons to having the work done on the outside or with freelance and you should gather information from other organizations in your area to support your position. Again, ITVA provides a yearly nationwide salary survey that is of great value in this area.

If the investment is to be a major one, sometimes it is best not to ask for everything at once. Structure your recommendations in easy-to-manage stages over short time periods. Budget each stage separately, with review points between stages so that management knows that it can stop at any time, if necessary, without being committed to a seemingly outrageous figure. A million and a half dollars does not seem like such a large sum when it is broken down into increments of $375,000.

Unless there is an overriding reason in a start-up situation, you probably should not recommend the purchase of production equipment until the organization is at least in its second year of program production. It will be enough of a job during the first twelve months to get management to okay the playback equipment, minimal staff allocations and use of outside production sources.

You will have the opportunity during this period to find out if you really need to buy equipment or if it is better to rent or continue to use outside production services. You may find your needs will change drastically once video is actually in operation in your organization. Why get stuck? Get management sold on the medium first. Then, after several successful programs, show how the organization might save money by letting you have your own production equipment, if you really think you need it.

During the start-up period you should recommend retaining an experienced outside producer/director to coordinate and supervise the production of your initial programs, if you are not an experienced video person yourself. These programs should be produced through outside facilities, which will give you a feel as to whether or not you want to have that capability in-house.

Management and The Tube 49

As we said earlier, you may want to suggest retaining an outside consultant to carry out a full-scale needs analysis or review the internally generated analysis and assist you in making your recommendations. The impartial third party usually has greater credibility.

One organization followed this path in developing video as a communications tool. In a memo to management, the communications manager pointed out that various units within the company were producing video programs on their own. In the past the Corporate Communications department had done one or two programs a year on a scattershot basis.

Playback equipment was rumored to be in place in a number of field locations. He requested authorization to look into the matter and see if there was a corporate-wide need for video programming on a regular basis and to find out whether or not an *ad hoc* network was in place. Permission was granted.

His subsequent report included a recommendation to bring in an outside consultant to carry out a needs analysis, since he felt he could not do it as quickly as he would like to. After an analysis of the programming produced in the past and its acceptance, interviews with potential video users and a canvas of all video playback and production equipment available in the field, a plan was outlined for the further development of video programming throughout the corporation confirming the manager's earlier conclusion.

If you are using video for the first time, the initial programs are crucial. They must be professional but not super-slick and their costs should fall within acceptable private television guidelines.

Because you want to involve management early, the temptation is to do a message from the CEO as the first show. It may be a mistake unless he is comfortable with the medium. If he is not, your entire plan may be in jeopardy if he looks bad.

Bill the first program as a "test" and pick a topic that has broad interest throughout the organization. If a company official is to appear on camera as "talent" be sure it is someone who is politically neutral. Or, you may do a test program for a specific division and then merchandise the results to the rest of the organization.

A word of caution on using any top manager as a spokesperson. Make sure, even if they are used to video and have undergone media training, you do a brief "test" before you start shooting the actual programming. No matter how sophisticated a presenter, he will want to correct any little thing that might later be embarrassing to him and to you.

Once management is sold on private television and is convinced that it's really doing a fine job, do not let up on your sales efforts. Don't sit back and tell everyone what a great job you're doing for them. Make doubly sure you are out asking people how you can help them with their training, marketing or communications problems. If you hear about a new project coming up, offer your help. Get in and sell video and your expertise. Don't wait until you get stuck with an impossible program idea or script and then have to try to make it into a program just because it has been "approved."

Chapter 6
Who Is Using Private Television?

Video Is Everywhere!

Meetings of most professional organizations now feature at least one or two speakers who will use video as a regular part of their presentations. Where once the organizers asked speakers if they would be using slides, they now ask what format tape they will be using.

Labor organizations are using video to inform their members of contract negotiations. A recent national meeting of the AFL-CIO was termed "a television driven convention" by the press because the union provided daily satellite transmission of convention highlights and preproduced tapes to 540 TV stations nationwide.

Video was used extensively by both parties during the United Airlines pilots' strike in 1985. The 5,200 striking pilots said that videocassettes and nationwide video teleconferences produced by the union helped them to maintain their solidarity. The airline countered by sending some 5,000 videocassettes to the homes of the striking pilots explaining the company's position.

"We're communicating the most direct way we can through VCRs, television and newspaper ads," said a union spokesperson. "About 40 or 50 years ago, we would have had mass rallies and soapbox speeches. But today it's possible to tailor specific messages to specific audiences such as families."

According to another media consultant involved in the dispute, "The video medium is the hottest way to communicate today. Pilots are using high-technology, so all of these devices just open up another level of communication that everyone is comfortable with."

At Allstate Insurance, video is at the core of every major new corporate program that is introduced throughout the company.

Safety meetings in most organizations today, regardless of size, use video as a part of their regular program. That doesn't mean just for a recreation of an accident or for role-playing either. Professionally produced motivational tapes get the safety message across sometimes even in a humorous vein.

Real estate offices in shopping malls offer video tours of available homes and properties. Executive search firms send videotapes of job candidates to personnel departments.

You can't get away from video even in the supermarket and department store, where a VCR brings you the latest in kitchen wizardry or designer fashions. Even the "Monsters of the Midway" have a promotion "music" video which plays endlessly in bars and restaurants in the Chicago area.

No one is forcing all of these organizations into video. It has come about naturally. Smart managers were the first to adopt private television for internal communications and training, and now see it as a necessity for an ever-increasing number of external uses as well.

Users by Size and Type of Organization

One of the interesting shifts we have noticed in the course of conducting the last three studies is the shift in the sizes of the organizations using video. Once it was said that only the big companies or large organizations could use video. Video then was a "big buck" item in anyone's budget. Maybe this was partially true when video first entered the corporate hallways. Equipment was more costly in those days and this limited the initial use of video by some small and medium size companies.

In the 1973 study, 41 percent of the users had over 25,000 employees. Today organizations of that size represent only 24 percent of the responses while organizations with less than 5,000 employees account for 40 percent, up from 19 percent in 1973. It is obvious that more and more small and medium-sized organizations are now enjoying the benefits of video. (Table 6-1)

What is apparent is that size has no longer anything to do with whether or not an organization is using private television. The primary reason that any organization uses video is because it does an effective job in communications and/or training. The same kinds of problems

TABLE 6-1
Private Television Survey Respondents by Number of Employees/Members
N = 247

Number of Employees/Members	Mfrs N = 86	Non-Ind N = 139	Med/Educ N = 8	Gov't N = 9	Other N = 5	% of Totals N = 247
Under 5,000	34.9%	38.9%	87.5%	77.8%	60.0%	40.9%
5,001-10,000	22.1	15.1	12.5	22.2	—	17.4
10,001-25,000	11.6	24.4	—	—	—	17.8
25,001-50,000	15.1	13.7	—	—	40.0	13.8
50,000-100,000	11.6	5.8	—	—	—	7.3
Over 100,000	4.7	2.1	—	—	—	2.8
	100.0%	100.0%	100.0%	100.0%	100.0%	100.0%

are faced by an organization with 1,500 members or employees in four or five locations as by a company with 100,000 employees spread throughout the world.

Typical Video User

Therefore, our profile of the typical video user remains as valid today as it was when originally outlined over ten years ago. The elements which fit almost every organization using video to some degree or another are:

- A highly diversified workforce with a population ranging from scientists and engineers, professional managers and skilled technicians to office and clerical personnel, semi-skilled and, in many cases, totally unskilled entry-level personnel.

- A geographically dispersed organization with plants, regional offices, international representatives and multi-divisional headquarters located throughout the country and sometimes in a growing number of overseas locations. Or, it may be a handful of plants, stores, service garages, sales offices or hospital buildings scattered over an industrial park or hospital campus, a single county or a major metropolitan area, but separated physically from the organization's headquarters.

- A multi-division/multi-market diversification. Different business units and differing markets often result in radically different orientations between divisions within a single organization. This applies as well to medical centers with trauma unit personnel dealing with different patients and procedures than cardiovascular surgery or pediatrics research units.

- Employees who want to know more about their organizations than just what is related to their jobs. They want to know more about the organization's goals, about its stand on the major issues that may be confronting it (e.g., takeover bids or doing business in troublesome foreign countries), and about their personal rights and company benefits.

- A management which believes that there is a direct relationhip between communication and the realization of organizational goals. This is not an afterthought in our profile. It is the key to the success of private television because without this commitment the entire effort is ineffective.

Years of Private Television Experience

The private television industry has not only come of age, but has grown a little gray around the temples. Over 75 percent of the video users responding to the survey have been using video for more than four years. That compares to 59 percent of the users in the same category in the 1981 study.

One-third of the respondents have been using video for over 10 years as can be seen in Table 6-2. In 1981 this figure was 17.2 percent.

The non-industrials which have been slightly more mature as users in the past surveys are now neck and neck with the manufacturers segment in proportionate longevity.

TABLE 6-2
Years of Private Television Experience
N = 247

Number of Years TV Used	Mfrs N=86	Non-Ind N=139	Med/Educ N=8	Gov't N=9	Other N=5	% of Totals N=247
Less than 1 year	1.2%	2.9%	—	—	—	2.0%
1 to 2 years	5.8	6.5	—	—	—	5.7
2 to 4 years	11.6	18.0	12.5	33.3	40.0	16.6
4 to 6 years	24.4	15.8	25.0	—	60.0	19.4
6 to 10 years	24.4	23.7	25.0	22.2	—	23.5
Over 10 years	32.6	33.1	37.5	44.5	—	32.8
	100.0%	100.0%	100.0%	100.0%	100.0%	100.0%

Chapter 7
Video Uses and Applications

The communications needs of all organizations are rising rapidly. So are their training needs. Several factors are behind this. First, corporate management (and this includes non-profit institutions as well) has finally recognized that communications is *vital* to the operation of their organizations. While management has always given lip service to communications along with motherhood, the flag and free enterprise, they rarely backed it up with any sincere commitment . . . or budget allocations.

Now management is discovering that communication is not only good for its corporate health, but that there is more of it to do than ever before. However, they frequently find that good intentions are not enough and that their organizations often lack the means or systems for effective communications.

In addition to the communications needs, the *training* needs at all levels throughout most organizations are becoming overwhelming as well. Basic skill training, generic and site-specific job training, supervisory skill training, management development, and safety are among the more urgent of these needs.

Because these needs are usually specific to individual locations, the training functions in most organizations are highly fragmented and dispersed throughout the organization. As a result, there frequently is little or no centralized coordination or planning to assist them. There is little or no pooling or sharing of needs and resources which leads to a duplication of effort among the many locations. Each staff or operating unit is usually on its own and individual trainers have told us that they often feel cut off and isolated. Consequently, many trainers keep "reinventing the wheel".

While at department levels, efforts are being made to develop generic skills, job and supervisory training programs that can be used at many different field locations, site-specific job and skill training needs are going largely unmet because of time and budget limitations.

The need for basic skills and job training is increasing at all field locations for a number of reasons. One is that work force reductions have necessitated cross-job training so that fewer people with more skills can do a greater variety of tasks. In addition, in many locations more sophisticated equipment and higher technologies (e.g., computers and microprocessors) require higher skills.

These factors, combined with a greater emphasis on individual worker productivity and the fact that newer employees are less prepared in basic skills, are placing heavy burdens on field trainers. Because of these pressures, from the top on down in many corporations there is a virtually unanimous feeling that *video should be used more extensively for communications and training*.

Just Another Tool

One of the barriers to a more effective use of video is the perception that there is only one kind of video and that it is an esoteric process requiring rooms full of expensive equipment and highly trained technicians. Certainly this was the case with data processing systems when they first entered the corporate environment and for many years it was also true for corporate television.

However, swift changes over the past two to three years have radically changed the nature of both technologies. Micro and mini computers have brought the power that once could be had only with giant mainframes to the individual user. Similarly, anyone can now create simple, good quality videotapes in color with low-cost, easy-to-use video equipment.

The important thing to remember is that neither personal computers nor home video systems will replace the larger systems in their respective technologies. Rather, they make possible myriads of *new uses and applications* that would be impractical, too costly or simply impossible for the larger, centralized systems to execute.

Many of these applications are site-specific and require no work beyond the initial recording. That is, no editing or duplication is required since the primary purpose is simple training and documentation and there is no need for a finished "program" with titles, graphics, music and so on.

Video Uses and Applications 59

There are literally dozens of these local video needs throughout many organizations that do not require the skill or expertise of a centralized video services staff.

When video first got started in organizations it was used as a fast, low-cost training aid. Role-playing and the recording of instructional chalk-talks or other presentations for repetitive use were the primary products. Editing was expensive and difficult, so little was done. Distribution was a problem because of the lack of format standards and equipment incompatibilities. Therefore, programs were usually shot and played back on the same machine in the same location.

With the advent on the videocassette, programs became more sophisticated and viewers' expectations rose. "Talking heads" were replaced or augmented by location shooting, flashy graphics, music and sound and professional editing techniques. Program formats and styles consciously and unconsciously copied what was seen on the broadcast networks, particularly on news programs. Video "programs" became television "shows".

But things are continuing to change. Over the last several years we have noted a decline in the full-length (20 minutes on the average) feature program. Where formerly it was an "event" for everyone to gather in a conference room periodically to watch a company-produced "TV show", video has now become a rather commonplace phenomenon in the work environment. As a result, viewer needs and expectations are changing.

The "program" as a concept is on the wane in favor of shorter, more frequent video "messages" or "memos", usually dealing with a single subject, that are incorporated into meetings or presentations along with other material. In most work locations, people want more video but supervisors are reluctant to "pull everyone off the job to watch a 20-minute program". Short video "modules" meet their needs better. This is covered in further detail in Chapter 8 ("Programming").

This is not to say that the concept of the video program is dead. Far from it. But the fully-produced video program is now perceived as not the only way to serve all video communications or training needs.

Video is just another tool to do a job. There should be no mystique attached to it. Requiring all video needs to be met by a centralized unit at headquarters is like requiring all screw drivers or typewriters to be kept in one location. Like all tools, video should be readily available to the user wherever he or she is located.

Training vs. Communications

It has become axiomatic that the use of video within an organization cannot be justified on the basis of a single application. Other audiovisual media such as slides, films, multi-media presentations and *ad hoc* teleconferences can be "one shot" in nature. You don't have to build an entire delivery system nor series of programs around them.

Video is always more successful when it is used to produce and deliver programs covering a *variety* of applications since it usually requires an organization to make two basic commitments. One is to invest in some form of playback or program distribution system. This nearly always requires a hefty financial outlay. The other commitment is to assign a person to function, even on a part-time basis, as an in-house *video producer*.

While video can be used for a number of different program applications, certainly it is not a solution for *all* communications and training needs. It must be considered in terms of its most important uses.

In the past when video or private television was mentioned, the application most people thought of was *training*. Blank looks crossed their faces when you started talking about *communications* applications. At that time, all private television programs were called *training programs*.

While video got its start in most organizations in the Training Department, its benefits soon spread to other areas and other uses and applications were developed, particularly for purely communications purposes.

Today most corporate video applications can be clearly divided between training and communications. Admittedly, there is a fine line between the two when certain program types are considered. Are *Employee Orientation* and *Benefits* programs training or communications? Take your choice. On the other hand, a videotape on how to dismantle a boiler feed pump clearly serves a *training* purpose, while an employee news program or a videotape reporting on the company's annual shareholders meeting can only be considered as *communications* applications.

It is easy to say: "What difference does it make?" Actually, a lot.

Communications and training programs are fundamentally different, although the equipment and the technical expertise used for each may be the same.

Video Uses and Applications

Communications programs generally originate at an organization's headquarters and are distributed widely throughout the organization. They have a greater sense of urgency to them since they are a form of "news". They usually have to be shown to their audiences more quickly than training programs. And, like broadcast "news" programs they are generally much more perishable and have a very short "shelf life."

Training programs more often than not originate in the field or at specific job locations and are not distributed widely. Less copies of training programs are required than for communications programs.

Communications programs are much broader in their objectives (sometimes to the point of almost being vague). Accordingly, results are much harder to measure. For example, how can you tell if a manager really understands his organization's position on an important regulatory matter when it may not be directly related to his immediate job? Yet it may be very important to the company that he be informed of the matter.

Training programs are much more specific in their objectives, particulary "skill" and "job" training programs. For example, a worker must know how to do a certain job or operate a specific piece of equipment. The training program is designed to do just that and if it is ineffective, the results are immediately apparent.

In many manufacturing operations, individual plant training directors produce many of their own videotapes. These are usually site specific (i.e., they train people at a given site how to operate machinery, fulfill orders, etc.). In these locations plants have their own cameras as well as videotape recorder/players. Production and distribution of these training tapes is generally confined to that location, except where a tape may be produced for several plants in one division which might have the same procedures or the same equipment or processes. The training group at headquarters may only act as an advisor to the plant or division training director.

Training programs require more advance planning and preparation and often take longer to produce. However, their "shelf life" is generally much longer than for communications programs. We know of some training tapes that are over 10 years old and are still going strong. The only reason they might be changed is to update them for cosmetic reasons . . . to change the clothing or hairstyles of the people involved.

The viewing environments for training and communications programs also differ. Training programs are generally viewed in classroom situations, many times with an instructor present who may use other audiovisual and printed materials in the session. Or, the videocassette may be viewed by an individual with other teaching aids.

Communications programs are generally shown in meetings which may be called specifically to view the videotape or which may use the videotaped subject as one item on an agenda. Some information programs are run in employee high-traffic areas for viewing at the employee's whim.

Finally, training is generally a *cost* associated with doing a specific job that relates to a profit-producing or cost-recovery activity. Communications is generally an *expense* associated with the general overhead of doing business and usually is not directly recoverable.

Applications In Ranking Order

Training programs continue to pay the freight in most video operations. Seven out of the top ten program categories in Table 7-1 are training applications.

The communications programs, which make up about 60 percent of the list of the video applications in the study questionnaire, are still growing, although not at the rapid rate they once were. Their ranking order in the charts has stablized (Table 7-2), with a few exceptions.

The top ranked video use in 1985 is *Skill Training* with 75 percent of the users producing programs in that category. It was also the top use in 1981 with 73 percent reporting skill training program production.

The next three in the 1985 top ten have moved up in the ranks as well as in the percentage of use over the past four years. In 1981 *Employee Orientation* was ranked third with 64 percent of the respondents using video for that purpose. Today, 73 percent produce orientation programs moving that application to second place.

Employee Information programming which ranked fifth in 1981 with 59 percent of the respondents, jumped to third place with 71 percent producing informational programs.

The biggest increase of the three categories is in *Employee Benefits* programming. Ranked tenth with only 49 percent of the respondents using video for the purpose in 1981, benefits has leaped to fourth place

TABLE 7-1

Current Video Applications In Ranking Order of Use by Total Respondents
N = 247

Video Application	% of Total Respondents
Skill Training	75.0%
Employee Orientation	73.0
Employee Information	71.0
Employee Benefits	67.0
Job Training	66.0
Management Communications	64.0
Safety/Health	63.0
Supervisory Training	60.0
Sales Training	60.0
Management Development	58.0
Product Demonstration	53.0
Sales Meetings	50.0
Community Relations	46.0
Employee News Programs	43.0
Annual Reports/Meetings	38.0
Professional Upgrading	33.0
Point-of-Sale	29.0
Economic Info/Education	22.0
Government/Labor Relations	21.0
Security Analyst Presentations	17.0

with 67 percent now employing the medium. A healthy increase of some 37 percent.

Two other categories which have increased in usage over the last four years have been *Safety/Health* and *Employee News Programs*. Although they both have moved up in rank only slightly, the percentage of their use has increased considerably. *Safety/Health* was in ninth place in 1981 with only 51 percent of the votes, today it is in seventh with 63 percent. *Employee News* has moved from sixteenth to fourteenth, and has experienced a growth of 56 percent.

Overall, the current use of video for every application is much heavier than it has been in the past. With the exception of the top ranked, *Skill Training*, which has always commanded over 70 percent of the responses, several applications over the last three studies have increased considerably in usage. These include: *Manangement*

TABLE 7-2

Current Video Applications In Ranking Order Compared With Uses in 1981

1985 Applications	1981 Applications
1. Skill Training	1. Skill Training
2. Employee Orientation	2. Job Training
3. Employee Information	3. Employee Orientation
4. Employee Benefits	4. Supervisory Training
5. Job Training	5. Employee Information
6. Management Communications	6. Management Communications
7. Safety/Health	7. Management Development
8. Supervisory Training	8. Sales Training
9. Sales Training	9. Safety/Health
10. Management Development	10. Employee Benefits
11. Product Demonstrations	11. Product Demonstrations
12. Sales Meetings	12. Sales Meetings
13. Community Relations	13. Community Relations
14. Employee News Programs	14. Proficiency Upgrading
15. Annual Reports/Meetings	15. Annual Reports/Meetings
16. Proficiency Upgrading	16. Employee News Programs
17. Point-of-Sale	17. Point-of-Sale
18. Economic Education/Info	18. Economic Education/Info
19. Gov't/Labor Relations	19. Gov't/Labor Relations
20. Security Analyst Presentations	20. Security Analyst Presentations

Communications, Product Demonstrations, Sales Meetings, Community Relations, Annual Reports/Meetings, and *Security Analyst Presentations*, as well as those mentioned previously — *Employee Information, Employee Benefits, Safety/Health* and *Employee News*.

Future Uses of Video

The top ranked *new programming application* for video in the next year will be *Employee News* according to the respondents who were asked to project their uses for the medium into the future. Thirteen percent of the respondents said they would begin doing this kind of programming in 1986 (Table 7-3).

Ranked as second in the future uses column are *Employee Orientation* and *Employee Information* and third place is held by *Employee Benefits* and *Management Development*. Coming in as fourth is *Management Communications*.

TABLE 7-3

**Future Video Applications In Ranking Order
of Probable Use by Total Respondents
N = 247**

Probable Application	% of Total Respondents
News Programs	13.4%
Employee Information	10.1
Employee Orientation	10.1
Employee Benefits	9.3
Management Development	9.3
Management Communications	8.9
Supervisory Training	8.5
Community Relations	8.1
Annual Reports/Meetings	7.7
Point-of-Sale	6.1
Sales Training	6.1
Proficiency Upgrading	6.1
Job Training	5.3
Safety/Health	5.3
Skill Training	4.9
Security Analyst Presentations	4.9
Economic Info/Education	4.5
Product Demonstrations	3.6
Sales Meetings	2.8
Labor/Gov't Relations	2.8

Over the last three studies these five uses of video have always remained at the top of the Future Use chart. They are all aimed at enhancing the informational level of the managers and employees at a most critical time in organizational history in our country.

As we have often said, due to its acceptance as a believable news medium in the home, video is also a believable medium in the corporate environment. Its use is perceived by employees and managers alike as a sign of an open climate of communications. The act of communicating is frequently more important than the message itself and video's inate impact intensifies this effect.

Chapter 8
Programming

Monkey Island

For as long as we have had "Private Television" — well over 15 years by last count — everbody has been using the term "talking heads" with a sneer. They usually are referring to an unimaginative video production that is guaranteed to put everyone to sleep because it is not peppered with splashy graphics, zippy location shots and dynamic special effects.

"My Gawd, can you believe it! A human face! With the lips moving! And words coming out!"

They almost make it sound like a Grade B horror thriller.

Our immediate reaction is to say, "Get serious! Most of the information you get everyday comes from some kind of *talking head* or another." Face-to-face communications always has been, and always will be, the preferred way of sending and receiving information between people . . . or as we said in an earlier chapter, "of getting a thought or idea out of one head and into another." Video is only the second best way.

If it wasn't for mankind's love affair with his own puss, there wouldn't be any television. Television is "Monkey Island"! If you have ever been to a zoo with a Monkey Island you will have a clear picture of what television is all about.

What are all of the monkeys on Monkey Island doing? Most of them are just sitting around watching what the other monkeys are doing

... which is chasing, fighting and/or making love, often all at the same time. Unlike us, the monkeys don't have to switch channels to go from one activity to another. They just sit and wait and something interesting usually happens right in front of them. Sounds just like an evening at home in front of the tube doesn't it?

Every living species shows a greater interest in its own kind than it does in any other life form. This is really why we watch television. It is our version of Monkey Island. There are chases, fights and a lot of love making ... often all at the same time.

Before television we did the same thing the monkeys do ... as a matter of fact we still do it ... hang out in the local piazza, town square or shopping mall. In the old city neighborhoods they put a pillow on the tenament window ledge and watch the kids and neighbors down in the street. Poor man's TV!

Those of us in communications have to face the fact that people would rather look at other people than anything else. This is most evident in broadcast television. In spite of the snide references to the "talking heads" the most popular televison shows are little else. Dan Rather and Tom Brokaw are "talking heads" and so is everyone on the Donahue and Merv Griffin shows.

This is particularly true of daytime network television with its heavy load of talk shows and soap operas. What do we see? Hour after hour of very tight close ups of faces, faces, faces — frowning, crying, gloating, sighing ... a lot of sighing ... and sustained meaningful reaction shots, usually accompanied by ominous musical "stingers."

There is a major difference between the way soap operas and televison action programs and theatrical films are shot. Mostly it is in the use of very tight close ups. Soaps use close ups for two reasons. One, the tighter the shot the less you have worry about the set. In the early days, before soaps became an important source of revenue, they were shot on shoestring budgets. Directors learned that by staying tight on the actors most of the time less money had to be spent on sets.

Second, the early TV screens were very small and resolution was poor. You needed to stay close just to get a good picture. When the producers found that the audience loved the tight closeups, they became a way of life. Today the sets are much more lavish, but the tight close ups remain because people want the intimacy of the faces.

However, if you used the same technique on a theatrical feature film, the results would be disastrous. Imagine a tight hair-to-chin closeup of a leading character on a giant Cinemascope screen. It would be overwhelming. Television directors learned this quickly a few years ago when they tried to use their regular techniques on large-screen, closed-circuit video teleconferences. The huge faces nearly drove the audiences out of their seats.

Most corporate video directors seem to be afraid to get the lens too close to the boss. We watched a batch of videotapes produced at one corporation and the closest the camera got to any of the managers was a very long "two-shot". You had to read the supers to be sure who it was. A new production was coming up featuring an interview with the chairman, a kindly but somewhat shy gentleman. We sent the director home for three days to watch soap operas and learn how to do tight head shots.

The set up on the day of the shoot was close and intimate and the interviewer was one of the communications managers the chairman liked and trusted. During the entire interview the bottom of the frame never got below his shoulders and the top was never more than an inch over his head. For the first time the man's natural warmth and gentleness came across for all to see. It became one of the most popular videotapes ever distributed in the company.

The Other End of the Horse

First and foremost, television is a *people* medium. It is best at conveying the communications associated with the human face. It is lousy as a means of readily communicating facts and data (except that which needs frequent updating, such as airport terminal departure schedules, stock quotations and so on). There are better ways of doing this than television. And by this we mean "television" in the *program* sense as opposed to "video" in the purely *display* sense.

In addition to the anguish and other heavy emotions generally portrayed on the soaps, "talking heads" are just great for communicating important company information. In fact, most people don't want it from any other source.

Every study we have conducted as well as those conducted by others say that employees want to hear company policy directly from the people who set it. They want to get key policy information directly from "the horse's mouth." Unfortunately, too often they get it from the other end of the horse.

The chief executives of organizations that use television for communications as well as for training instinctively know the value of face-to-face communications. As one chairman put it: "Television enables me to project my personality and my views throughout the organization." As it turned out he was absolutely right. He had an obnoxious and abrasive personality which television communicated quite well. More people got to hate him than would have without television and the board eventually fired him.

The trick, of course, is to come off as the "horse's mouth" without being a "horse's ass." It is the other end of the horse that has been giving "talking heads" such a bad name over the years, not the concept itself.

Actually, there is less of a problem in an organization with hourly or non-supervisory workers appearing on camera when producing a corporate information program than with management people. We have found that the "talking heads" on the shop floor are succinct and to the point. In many cases they are far more credible than their bosses. They are usually good on the first take and require the least amount of editing. Managers, on the other hand, tend to be more self-conscious and stiff. They often back into statements with endless qualifications and generalities and usually are murder to try to edit.

With middle management you can get them to do it over until it comes out usable. With top management you have to be more careful. They are always in a rush and think they can do what they have to in front of the camera in one take with no rehearsal and be gone.

Furthermore, they are not used to being told what to do, particularly by people who work for them. On the other hand, they are very aware (or should be) that when they are paying for professional expertise they should use it. If you _are_ a professional and continue to demonstrate it, they will trust you and follow your direction. This, of course, is vital in keeping the right end of the horse pointed toward the camera at all times.

Being afraid to correct the boss when he or she is in front of a camera is probably about the worst mistake a corporate video producer can make. Our favorite story concerns a cigar-chomping chief executive who made his first videotape. Everyone was afraid to tell him that his nicotine-stained dentures looked terrible. When they showed him the completed program, he saw how badly he looked and hit the roof. Several thousand dollars and a new set of dentures later they reshot the show . . . with a new director.

Getting top management to take direction so that they look like the right end of the horse is generally not difficult when you are dealing with them on a one- on-one basis. The problems start with the phalanx of anxious upper-middle managers who tend to surround the chief like a belt of armor plating. A technique one director uses in situations like this is to waste a take, fake an equipment problem and review the tape with the talent during the phony downtime. This is easier and more effective than trying to make corrective suggestions during a preliminary runthrough or rehearsal with everyone standing around looking nervous.

In spite of the air of command associated with most top executives, studies have found that a large number of them are basically shy, insecure people. They need reassurance. And they need it from someone whose judgment they can trust, other than their coterie of "yes" men.

One board chairman recently asked us to review the videotapes he had appeared on and tell him what was both good and bad about them. "I've spent a lot of time watching them," he said. "In general I do all right. But there are a lot of things I don't like. I don't know why I don't like them or what I should be doing differently. None of my staff has been much help. They all tell me how great I look. That's bullshit!"

There are a lot of consultants in the business of coaching executives on how to appear on camera and to deal with the news media. Some are former newsmen who concentrate on training executives to be able to respond to the media. Others concentrate primarily on personal presentation techniques. However, looking good on camera is not necessarily the same as looking good behind a podium making a speech. Different techniques — and different coaches —may be required. Virtually all of them do an excellent job. You just have to pick the right one needed for the specific assignment.

Keeping the right end of the horse aimed at the camera is now more important than ever before in private television with the increasing use of non-program, video memos for announcements and other applications, as we shall see later in this chapter.

Levels of Production

The number of video programs produced by the typical Private Television user/producer has risen considerably since the last study. In 1980, the median number of programs produced per organization was 18, down slightly from the median of 21 reported in 1977.

In the last five years, however, the median has climbed back up to 23 in 1984 and last year it increased to 25. This is an increase of 39 percent over 1980. User predictions for program output in 1986 are even higher! The median number of programs the respondents said they would produce this year is 31, *a 24 percent increase over the preceding year!*

This works out to approximately *one program every four days* of active production time *per video operation*, based on the 1600 production hours per year described in Chapter 9. It is a hefty work load, no matter what!

In this study, for the first time we asked the respondents to break down their production totals by *communications* and *training* programs. Looking at the breakdown in Table 8-1, we see that the median number of *communication* programs has risen to 12 in 1985 with a further increase to 15 for 1986. The same increase holds true for *training* programs as well, with the median rising to 15 in 1986 over the previous year's median of 12.

TABLE 8-1
Median Number of Programs Produced By Total Respondents

	1984 N = 230	1985 N = 235	1986 N = 248
Total Programs	23	25	31
Communications	10	12	15
Training	10	12	15

The level of programming has shifted during the last four years at both ends of the spectrum as well as in the middle range (Table 8-2). Fewer organizations are producing 10 or less programs each year. In 1981, 30 per cent of the respondents fell into that range, while today only 18 percent produce less than 10 programs a year, a decrease of 39 percent.

At the other end of the scale we find that while 16.6 percent of the respondents produced 61 or more programs in the last study, 19 percent produced that number in 1985. This is an increase of nearly 15 percent.

TABLE 8-2

Number of Private Television Programs Produced In 1985

N = 235

No. of Programs Produced in 1985	% of Total
1 to 4	6.8%
5 to 10	11.5
11 to 20	23.8
21 to 40	22.1
41 to 60	15.8
61 to 100	12.8
Over 100	7.2
	100.0%

We see some interesting trends behind the figures. As we will see later in this chapter, *a program is not always a "program."* Second, the steadily rising level of program output seems to reflect the increasing level of professionalism and competence on the part of the average video person. That is, more programming per person is being produced than ever before. In addition there is also the contribution being made by the availability of better equipment, particularly in the post-production area.

Even though the numbers are increasing, we find that the approach to production is also more realistic. With the advent of the videocassette in the early 1970s, we often encountered "run-away-video." That is, video was used to try to solve just about every communications and training problem that could be found, whether it was suitable for that particular purpose or not.

The tendency now is to use video only for applications for which it is best suited. More video people seem to be asking the question: "How do you know you need a *videotape*?" when they are approached by an eager in-house client who has a "great video solution" to a poorly defined communications or training problem.

On the other hand, real problems can develop if staffs and facilities become overextended. There are only so many projects that can be completed while maintaining a sufficiently high level of quality in any production department. With the median number of programs at an all-time high, and the demands for quality increasing, one wonders how an average video staff of three (see Chapter 13 — "Organization and Staffing'') can turn out a program every week without something or someone suffering in the process.

As one video manager told us, the real limitation on production is the amount of effective management time that can be applied to any one project, not the numbers of people or items of equipment which can be added incrementally. "When I am stretched too thin," he said, "everything suffers. Morale goes all to hell and the program quality plummets."

Corporate video is no longer a novelty and the demands for quality productions are continuing. Yesterday's "home movie" standards are no longer acceptable to audiences who are used to increasingly dazzling production values on their home screens.

Video Memos

As we said in Chapter 7 ("Video Uses and Applications"), in the last three years we have noticed a trend in some organizations to produce short, single topic "video memos" or messages that are turned out quickly with a minimum of editing, titles, music, graphics, and so forth.

Generally these "memos" are communications oriented, e.g., a "talking head" announcing a new policy which can be used as a meeting opener or an organizational change in personnel which doesn't need a great deal of explanation. These "non-programs" are produced instead of or in addition to regular programs that have some or all of the traditional "program" production values.

Fifty-eight percent of the survey respondents produce "video memos" of this type in addition to their regular programs. They also report that 21 percent of their productions are such non-programs. We see this trend continuing because these shorter modules are better suited to viewing in field locations as a part of a local meeting or presentation than are longer, more fully produced programs.

Program Sources

There are three sources of programming for most video users:

- In-house production
- Outside production
- Published programs

Programming

As indicated in Table 9-1 in Chapter 9 ("Production"), only 31 percent of the respondents produced all of their video programming solely through in-house resources. Two-thirds said they used a combination of in-house and outside resources. Only four percent said they did everything through outside services, a surprisingly low figure.

Published Programs

Purchases of published programs have decreased slightly in the last four years. In 1981, 46 percent of the respondents purchased a median of 10 programs. Today, 42 percent of the users reported purchasing a median of 8 published programs.

The leading subject area of published programs purchased is *Skill Training*, while the second is *Management Development*.

Of those video users purchasing published programs, 24 percent said they do the selection and buying of the programs, while 76 percent report that they help (presumably the program's end-user) with the selection and buying.

Length of Programs

At last organizations using video have realized that you cannot hold the viewer's attention for endless periods of time.

The median length of video programs is 17 minutes, with 93 percent of the respondents replying they produce programs 30 minutes or shorter in length (Table 8-3).

It is interesting to note that in 1981, 24 percent of the respondents reported producing programs of 45 to 70 minutes in length, while 13 percent reported average lengths of over one hour. Today those two categories combined total only 3.1 percent of the respondents.

This again reflects the overall trend toward shorter, non-program video modules.

In-House Clients For Video Programming

The line-up of in-house clients for whom the video unit produces programming remains the same as it has been in the past (Table 8-4). Personnel and Training are always the number one ranked video unit for whom programming is produced. This stands to reason since

Training uses are at the top in the applications listings cited in Chapter 7 ("Video Uses and Applications").

Corporate Communications is a close second, reflecting the growing amount of communications programming being produced by most user organizations.

It is interesting to note that 62.1 percent of the respondents say that Top Management is a production client. Obviously, top management which appears on video in 91 percent of the respondents' organizations, is a prime originator of video programming as well.

TABLE 8-3

Average Running Length of Programs Produced
N = 232

Average Length	% of Total Respondents
Less than 10 minutes	13.8%
10-20 minutes	53.4
21-30 minutes	26.3
31-45 minutes	3.4
46-50 minutes	2.2
Over 60 minutes	0.9
	100.0%

TABLE 8-4

Organizational Unit For Whom Video Unit Produces Programs Each Year In Ranking Order

Organizational Unit	% of Total
Personnel/Training	77.7%
Corporate Communications	70.5
Top Management	62.1
Marketing/Sales/Advertising	46.1
Line Divisions	34.4
Other (Patient Ed., etc.)	33.9
Legal	14.3
EDP/Systems	10.3
Treasury/Financial	8.9

(More than one answer may apply for each respondent.)

9
Production

In the past, the lines between in-house and outside production were clearly defined. In the 1973 study, 93 percent of the video users produced all or most of their programs in-house, while only 20 percent produced programs through outside resources.

In the late Seventies the lines became blurred. More programming was produced using outside resources as they became more plentiful. By the 1981 study we found that 83 percent of the users reported producing programs in-house while 60 percent claimed use of outside production sources.

This time we asked more specific questions regarding in-house and outside production and found that although 96 percent of the respondents produce programs in-house, only 31 percent say that they use no outside resources (Table 9-1).

TABLE 9-1
Source of Private Television Programs
N = 229

Program Source	% of Total
Produce Programs Only In-House	31.0%
Produce Programs Only Through Outside Sources	3.9
Produce Programs In-House with Outside Sources	65.1
Purchase Published Programs	41.9

(More than one program source may apply for each respondent.)

We also find that the 69 percent of the respondents use outside production sources. This is an increase of 9 percent over 1981. Of those using outside sources, 4 report this as an exclusive source.

Use of in-house and outside sources is no longer an "either/or" situation. Today, many organizations own only portions of the equipment and facilities needed to produce programs and rely on outside services for the the remainder. Furthermore, many firms with fully developed in-house facilities use outside facilities to do entire productions or will use specialized services for program completion.

Investment in In-House Production Equipment

The total investment in in-house production equipment and facilities has risen since the last report. In 1981, 28.1 percent of the users had over $100,000 invested in production equipment. Table 9-2 shows that in 1985, 43.8 percent have that amount invested, *an increase of 56 percent* in five years. Today's median level of investment is $83,500.

The primary production equipment owned by the respondents are ENG cameras, with 87 percent reporting ownership of these units. A large number said they used their ENG cameras for studio production as well. In addition, 44 percent of the respondents said they also owned studio cameras.

Investment in in-house production equipment of less than $50,000 showed the most significant drop on the chart. In 1981, 52.4 percent of the respondents fell into this category, while today only a third have this much invested.

The total number of respondents having production equipment and facilities has risen slightly since 1981. Today, 94 percent of the respondents have production equipment and facilities. In 1981 only 89 percent reported owning their own equipment.

Also of interest in Table 9-2 is the fact that with one exception in the government category, all non-corporate investments are valued at less than $250,000. Overall, some 26 percent of the respondents reported investment figures higher than a quarter million dollars. These are all corporate facilities.

Once again we find that the non-industrial sector has a slightly greater commitment to production equipment and facilities than does the manufacturing group. Many of these are old-line video operations which have maintained significant investments in broadcast quality equipment.

TABLE 9-2
Total Investment In-House Production Equipment
N = 224

Total Investment	Mfrs N=76	Non-Ind N=129	Med/Ed N=7	Gov't N=9	Other N=3	% of Total N=224
Under $10,000	6.6%	7.8%	—	—	33.3%	7.1%
$10,001–$25,000	19.7	9.3	28.6	22.2	—	13.8
$25,001-$50,000	13.1	18.6	14.3	11.1	33.3	16.5
$50,001-$100,000	14.5	19.4	42.8	33.3	—	18.8
$100,001-$250,000	18.4	16.3	14.3	22.3	33.4	17.4
$250,001-$500,000	14.5	13.2	—	11.1	—	13.0
$500,001-$1 Million	5.3	7.7	—	—	—	6.3
Over $1 Million	7.9	7.7	—	—	—	7.1
	100.0%	100.0%	100.0%	100.0%	100.0%	100.0%

Investment in Post-Production Equipment

Eighty-nine percent of the respondents said that they had in-house post-production equipment and facilities, showing a rise of some 10 percent over the 1981 respondents.

Table 9-3 tells us that two-thirds of the video users have over $50,000 invested in post-production facilities, with a median investment of $78,500. In 1981 only 44.3 percent had that sum invested in post-production equipment.

It is interesting to note that the manufacturing sector has a slightly higher commitment to in-house post-production equipment than does the non-industrial group of respondents.

Location Shooting

In 1977, location shooting was only a trend because ENG/EFP equipment was relatively new in the corporate world. Yet, three-quarters of the respondents said they shot a median of 39 percent of their programs on location. In the 1981 study, 84 percent of the respondents shot a median of 68 percent of their programs on location.

The numbers continue to rise. In 1985, 95.1 percent of the video user/producers reported that they shot a median of 72 percent of their programs on location either with in-house gear or using outside sources. It is interesting to note once again at this point that 87 percent of the respondents reported owning ENG cameras and this certainly accounts for a great many of the programs produced on location.

Size of In-House Studio or Production Area

The ownership of studio or fixed production areas remains roughly the same as in 1981 with 74.8 percent of the video users responding that they have such an area. The same percentage of respondents claimed studio space in 1981.

However, the size of the studio or fixed production space continues to drop, further proving the point made in 1981 that smaller was better. With the increasing use of ENG/EFP equipment and the fact that so much of the production is done "on location", there is no need for an immense 2,000 sq. ft. studio to house portable gear. In most cases, you merely need someplace to set up to do an insert, a product shot or other segment that can't be done in someone's office or at a field location.

TABLE 9-3
Total Investment In In-House Post Production and Editing Equipment
N=211

Total Investment	Mfrs N=73	Non-Ind N=119	Med/Ed N=8	Gov't N=9	Other N=2	% of Total N=211
Under $10,000	8.2%	4.2%	—	11.1%	—	5.7%
$10,001-$25,000	19.2	15.1	37.5	22.2	50.0	18.0
$25,001-$50,000	9.6	17.6	—	22.2	—	14.2
$50,001-$100,000	17.8	20.2	50.0	33.4	—	20.9
$100,001-$250,000	24.7	20.2	12.5	—	50.0	20.9
$250,001-$500,000	6.8	11.8	—	—	—	9.0
$500,000-$1 Million	9.6	8.4	—	11.1	—	8.5
Over $1 Million	4.1	2.5	—	—	—	2.8
	100.0%	100.0%	100.0%	100.0%	100.0%	100.0%

But there is a caveat. In some organizations a large, fixed studio is a necessity. If you are located in an inexcessible place with no facilities nearby for those occasions when you need to videotape your CEO for important policy changes or the annual employee report, then you may need a studio. Or, if you do many intricate product demonstrations and are located in the middle of nowhere, again you may need a studio. A third reason for studio facilities in-house may be the weekly production of a management program on highly confidential material which you do not wish to share with an outside facility.

Later on in this chapter we will cover the use of in-house versus outside services to a greater degree.

Since 1977, we have seen the size of the studio space fall. At that time one-third of the respondents had studios under 600 square feet, another third reported studios of 600 to 1,250 square feet, and the final third claimed space of over 1,250 square feet.

By 1981, those with over 1,250 square feet fell to 28 percent. Forty-four percent reported the mid-range sized studios of 600 to 1,250 square feet. The remaining 28 percent said their studio or fixed space fell under 600 square feet.

In 1985, the shift has been slight, but still downward, in studio size with 76 percent of the respondents now reporting space of under 1,250 square feet (Table 9-4). The bulge in the mid-range is higher with 46 percent now falling into the 600 to 1,250 square feet category.

TABLE 9-4
Size of In-House Studio
Or Fixed Production Area
N = 181

Size of Area	% of Total
Under 600 square feet	30.0%
600 to 1,250 square feet	46.0
Over 1,250 square feet	24.0
	100.0%

Location of In-House Studio or Production Area

The location of the studio or fixed production area has shifted back to organizational headquarters once again. In 1977, two-thirds of the respondents claimed organizational headquarters as home for their

Production

studio. In 1981, the figure dropped to 48 percent. Now, it has risen back to 60 percent (Table 9-5).

TABLE 9-5

Location of Studio Or Fixed Production Area
N = 182

Location of Area	% of Total
Organization headquarters	60.4%
Training Center	13.8
AV Media Center	19.8
Other	6.0
	100.0%

There is a logical explanation for this. With over 70 percent of the programs now shot on location, the use of portable ENG cameras, the "studio" smaller in size, it stands to reason that the area itself is probably installed in some former conference room near to the Corporate Communications Department and is not a major piece of dedicated real estate.

Training centers and audiovisual media centers — usually not located in headquarters — no longer seem to be the prime homes for video studios. Only one-third of the respondents reported being located in such an area. This is a drop of 32 percent from the 1981.

Production Formats

For the first time, respondents were asked questions regarding their production and mastering formats. The "workhorse" production format of private television is still the 3/4-inch U-format cassette. Nearly nine out of ten producers continue to use this for their primary production (Table 9-6).

TABLE 9-6

Videotape Production Formats
N = 244

Format Used	% of Total
3/4-in. U-Matic	88.1%
1/2-in. Beta (including Betacam)	17.2
1/2-in. VHS (including "M" Format)	14.8
1-in. Type "C"	3.1

(More than one answer may apply to each respondent.)

Beta and VHS are little used for primary production, only 17.2 and 14.8 percent respectively. One-inch type "C" barely makes a dent as a production format in this market with only 3.1 percent of the respondents claiming its use for primary production.

Mastering Formats

The U-format 3/4-inch cassette "workhorse" also dominates the post production area as a mastering format. Nearly two-thirds (62.3 percent) of the respondents master on 3/4-inch videocassettes for this purpose. Some 32 percent of the respondents do most of their mastering on 1-inch Type "C", with the remaining six percent using either 1/2-inch VHS or Beta to create program masters.

The New 8 MM Format

Each time we do a new study of the private television industry, by an accident in timing a major new video format is also introduced. In 1973, the 3/4-inch U-Matic videocassette was just getting started. In 1977, the new half-inch Beta and VHS formats came on the scene. In 1981, the videodisc was capturing everyone's attention.

Now we are being offered a new format — 8 mm videotape in a cassette small enough to be slipped easily into a shirt pocket for a camera/recorder no bigger than a paperback book.

Once again the questions are being raised. Do we really *need* a new format? What about compatibility? How can anything that small be any good? (The same thing they said at various times about the one-inch, three-quarter-inch and half-inch formats, by the way.) We have too much invested in other formats, how can this help us?

In 1977, *nobody* could see a use for half-inch cassettes in corporate video. Yet three years later virtually all new distribution networks and the expansion of most older ones were based on one of the two half-inch systems.

It took one or two major corporate applications to gain credibility for the new formats then everyone followed suit. Much of the same thing is happening again today with 8 mm. In early March 1986, Prudential Insurance, one of the oldest and most developed video users, announced the purchase of 1300 Sony 8 mm units for use by their field agents. The uses will be for applications that it would be impractical to have met by the established corporate video operation. It will be a case of not "instead of" but "in addition to" current video uses.

Production

We predict that 8 mm video has a bright future in the organizational environment. In addition to the benefits of lightness and extraordinary small size, it offers picture quality as good as if not better than half-inch. Footage recorded on 8 mm can be "bucked up" to any other format with little or no generation loss. In addition, the format is supported by more than a hundred manufacturers world-wide.

Post-Production Equipment

When asked what types of post-production and editing equipment they owned, 65 percent of the respondents said they own off-line editing units, while 70 percent claim they own post-production and completion systems. Seventy percent also have duplicating equipment of their own.

A closer look at the investment in post-production equipment also shows that there is a relationship between years in video and size of investment. Over three-quarters of the respondents having over $100,000 invested have been involved in video for six or more years.

Supplemental Outside Services

The use of outside services for production and post-production by the private television industry first seen as a growth area in 1977, continues to expand.

The single largest area of growth is in duplication with 80 percent of the respondents using outside resources to make multiple copies of their programs. (Table 9-7). The use of outside duplication services was reported by 50 percent of the respondents in 1981 and in 1977.

Another growth area in supplemental services has been program concepts and scriptwriting. In 1981 only one-third of the respondents said they used outside resources to perform those tasks. Today, 57.3 percent use such pre- production services, a growth of over 70 percent.

In this study, off-line editing and final editing/post-production was broken down in a different manner, therefore making comparisons to past studies difficult. However, of those who said they go outside for final editing, 79 percent in 1985 far exceeds the 44 percent reporting use of outside editing and post-production facilities in the last study.

Once again, those who have principal production equipment are using outside services for production services as well. Only 6 percent reported having no production equipment, yet 29 percent of the

TABLE 9-7

**Supplemental Services Used
(In Ranking Order of Use)
N = 157**

Supplemental Services	% of Total Using Services
Duplication	80.1%
Final Edit/Post Production	79.0
Program Concepts and Scriptwriting (Pre-Production)	57.3
Off-Line Screening and Editing (Pre-Post Production)	39.5
Principal Production (Shooting)	28.7

(More than one answer may apply for each respondent.)

respondents say they use outside principal production services and facilities. Therefore, the cross over of those who have equipment and still use outside services continues as was first reported in 1981.

In-House Studios vs. Outside Production Services

The question of in-house vs. outside production services will always be debated among corporate video users. The issue can never really be resolved, nor should it. The needs of each organization using video for communications and training are unique to that organization. Doing everything outside may be just right for one company, while its next door neighbor may be quite justified in doing almost all of its production in-house.

There are many factors involved in deciding whether to use outside services and suppliers or to develop an in-house capability to handle the same work. These include:

- *Speed* — Can the work be performed faster outside or in-house when tight deadlines are involved?

- *Cost* — When all costs are considered where will it be least expensive?

- *Quality* — Where will the best quality, in relation to cost, be achieved?

- *Convenience* — Where inclusion of products or services and/or facilities or participation by other organizational personnel is required which will cause the least disruption of operations and schedules?

- *Confidentiality* — Is the subject matter confidential or too sensitive for outsiders to handle?

- *Peak Loads* — At periods of peak work loads, which will be able to meet deadlines within budgets while maintaining quality?

- *Specialty Services* — Is a special expertise or technical operation needed that is not normally an in-house capability?

- *Familiarity* — Is a special knowledge of organizational operations, policies, products or services and/or procedures required?

- *Availability* — Are the required services available through outside suppliers for practical utilization or is in-house the only way they can be provided?

Putting it another way, *will the total annual cost of an in-house capability be lower than the total annual cost of purchasing the same services on the outside* when all other influencing factors are equal or not a consideration? The same philosophy applies to people as well as to equipment and facilities.

Nathan Sambul, editor of "The Handbook of Private Television" (McGraw-Hill, 1982), cites the "Four Cs" as the justification for the fully developed in-house facility that he built and managed at Merrill Lynch — *Convenience, Confidentiality, Cost and Control.*

While *cost* is usually given as the principal reason for an in-house operation, it is often used as a cover for another equally important reason that many companies are afraid to admit: *convenience*. According to Sambul, "If the chief executive is an active video user, he's not going to want to travel more than four floors in an elevator. Forget the limo to the studio uptown if he does more than one production."

Confidentiality is another reason Sambul cites for Merrill Lynch producing all of its programs in-house. This also is a key justification for IBM's studio in White Plains. If the protection of proprietary company information is a major concern, then an in-house facility is certainly in order.

Cost and *control*, however, are arguable points. No matter where a program is produced, the producer should have "control" of the project, whatever that all-encompassing term may mean.

To some, control may mean having a familiarity with the content that an outsider could not be expected to have. To others, it may mean the care and stroking of fragile executive egos on camera. And so on.

Frankly, we believe that knowing what you are doing gives you adequate control, wherever you produce. In some cases, we have seen in-house producers have greater control of their productions in outside studios where they are free from interference by company watchdogs and hangers-on.

The *cost* question can be argued endlessly, however. Creative accounting will make a convincing case in either direction, depending on what costs you choose to include or ignore. Allstate Insurance, for example, feels that if all of its shared corporate overhead costs are included in the cost analysis of its in-house productions, the productions would be more expensive that if they were done in a major Chicago video production house. "After all," says Steve Mulligan, Allstate's director of Audiovisual Services. "how many production houses have to include five acres of landscaping and a lake full of Canadian geese in their rate cards?"

Where the discussion of in-house vs. outside production really bogs down is over the question of "the studio". Many corporate people assume that if they're going to make video programs, a studio is required. It's a part of the video culture we've grown up with and it's a tough mental habit to break. One client recently told us they were about to produce a series of employee news shows and needed a studio to do it. "Why?", we asked. "Aren't you going to shoot most of the stories on location?" "Sure, but we need the studio for the anchor person." (Strangled groan!)

Studios get in the way of effective video production in many organizations. We told one major manufacturer to increase its use of video throughout the company by closing its studio. The facility was an albatross. The feeling was that it had to be used for all programs, but it was capable of only the simplest productions. Everything came out looking the same.

Studios were needed in the early days of television for two basic reasons. One, equipment was heavy and couldn't be moved around easily; it needed a permanent home. Location shooting was usually

Production

done on film. Second, sophisticated editing systems didn't exist, so programs were put together by switching live among several cameras during production. In addition, cameras weren't very sensitive, so complex lighting grids were needed not only to flood the sets with light, but to provide some semblance of depth to an otherwise flat scene through multiple key and back lighting set-ups.

A three-camera switched studio production is highly labor intensive. Approximately 10-12 people are needed, all performing specialized functions in perfect synchronization with each other during the full course of the production to the total exclusion of all other activities. If taping stops to reset lights, balance cameras, fix a piece of equipment or whatever, their time is lost during the wait. Most organizations cannot afford this kind of staffing on a full time basis. Many that do have production studios supplement a small core of insiders with outside freelances during a production.

Then there is the intangible, but often substantial cost of the management time that must be devoted to planning, budgeting and supervising, plus the problems that go along with the running of both people and equipment. All of the time devoted to administration is time taken away from getting the primary assignment, namely video production, completed.

But staffing is only one of the problems. One cynic recently described an in-house production studio as a hollow void in a building that video managers keep trying to pack full of money. It is illustrated by the following true story.

A young video manager came to his boss to tell him he needed a new piece of equipment. "How much?" the boss asked. "Only $5,000," replied the VM. "Is that *all* you need?" "Well, no, not exactly," came the reply. "A couple of things have to go with it to make it really work right with our system." "Like what?" asked the boss, feeling an icy grip in his stomach. "Well, we should have (blah) and to make that work we need (another blah) and then a (micro-blah), plus the connectors." "What's all that come to?" his boss asked in a very even tone. "Ah . . . we should be able to do it all for $55,000."

In a multi-camera production studio, there is no such thing as a stand-alone piece of equipment. Everything has to be hooked to something else, usually something costing twice as much. To the best of our knowledge, no in-house video studio has ever been finished. All are in a perpetual state of upgrading, conversion, expansion and so on. It is a never-ending process.

One reason is a sincere desire to produce ever better programs with no increase is manpower. (Ironically, many companies will buy a $50,000 piece of equipment before hiring an $18,000 staff person. The reason is that this person will end up costing the company approximately $350,000 in salary, benefits, raises and general overhead over a 10-year period. Also, there are no investment tax credits and amortization on people.)

The other reason is "run-away technology". Bells and whistles are proliferating faster than we can ring or blow them. Obsolescence precedes delivery. The head of a major New York production studio told us he ordered an $85,000 piece of equipment at NAB. Delivery was scheduled 10 months later. The week before the equipment arrived he received the brochure for the new model. Less than a year later he had to spend $35,000 to get the unit upgraded in order to stay competitive in his market.

Few in-house facilities can afford that kind of tail chasing. Even if they can, this is not the way they should be allocating corporate resources unless a positive impact on the bottom line can be clearly shown.

With no disparagement of the talents and experience of the in-house producer — many of whom have come from the world of professional television production — an in-house production can rarely match the technical quality of an equivalent program produced at an outside studio simply because the deck is usually stacked against it financially. It is virtually impossible to get the same production quality from a $500,000 in-house studio and one costing over five times that amount. Talent and creativity will only go so far.

Furthermore, the boss is going to be looking at an in-house production with the same pair of eyes he uses at night to watch a network special. Unconsciously, he expects the same technical quality from the non-broadcast product.

Many organizations with in-house facilities are also big users of outside services, as we have shown. "I just love companies with in-house studios — they're my best customers," says Robert Henderson, head of Windsor Total Video in New York City.

A good video manager realizes that he or she can't do it all and will go outside for specialty services that would be impractical to maintain in-house. Further, as video needs grow within the organization, there are bound to be peak periods when the in-house facility and staff are

Production

fully committed and a program still needs to be gotten out under a deadline. This is why commercial production/post-production houses cultivate in-house video people.

How far you can go with an in-house production "capability" — studio or otherwise — is, of course, dependent upon what services are available to you in your local area. In most major urban centers, production facilities abound. But if you're smack in the middle of a central state's soybean field with nothing more than more soybeans for hundreds of miles in all directions, then you're going to have to be self-sufficient.

Today the flat studio look is "out" and the *live* location look is "in". Cameras and recorders can go everywhere and return excellent pictures under what were once considered impossible conditions.

Single-camera film-style shooting is producing better composed and better lighted pictures. Switching errors are eliminated and talent goofs are more easily covered. The old film saying, "Don't worry, we'll fix it in post" is now true in video. Sophisticated editing systems now provide greater flexibility and visual interest in a program.

Today, most commercial production facilities do not offer the kinds of fully equipped studios they once did. The studio control room has been converted into a CMX editing suite and the heavy studio cameras are being replaced by general purpose ENG/EFP cameras that are at home anywhere. The studio is "just another location" and probably is more like a bare motion picture sound stage than the television studio as we've known it in the past.

Top management likes the new video better, too. More than once in the past we have been admonished about cluttering the CEO's office with equipment and running bundles of cables along plushly carpeted executive hallways when shooting management fauna in its natural jungle habitat.

Location shooting makes management more credible. Frank Perdue surrounded by hundreds of tiny naked yellow bodies and Frank Borman stalking down the aisles of a 727 are very believable. How much would you trust them delivering the same pitch in front of a potted plant in a studio?

Chapter 10
Distribution

Are You Suffering From "Network Narcosis"?

In our search for excellence, we all find ourselves falling victim of self-help quizzes. Here's one for the video network manager. It tests the health of your distribution network and your management acumen.

Answer the questions honestly and we'll help you analyze the score.
1. How long has your video network been up and running?

   a. ( ) less than 2 years
   b. ( ) 3-4 years
   c. ( ) 5 years or more

2. How often do you update your Network Field Coordinators' mailing list?

   a. ( ) every 6 months
   b. ( ) once a year
   c. ( ) every 2 years
   d. ( ) never

(Many organizations have officially designated *communications coordinators* in field locations who receive and set up the viewing of video tapes sent from headquarters. This is usually in addition to handling other communications duties. Others merely send programs to the local plant or office manager who coordinates the showings. For this quiz "Network Field Coordinator" means anyone who gets the tape to show to others.)

3. When was the last time you personally spoke with most (or at least a representative sampling) of the Field Coordinators responsible for showing your tapes?
   a. ( ) last week
   b. ( ) within the last year
   c. ( ) about 2 or 3 years ago

4. When was the last time you visited a representative sample of your field locations to check their viewing facilities and to review their videotape promotion/scheduling procedures?
   a. ( ) last week
   b. ( ) within the last year
   c. ( ) 2 or 3 years ago
   d. ( ) never

5. How often do you get feedback forms from your Field Coordinators telling you that everyone likes your programs "just fine".
   a. ( ) never
   b. ( ) once in a while
   c. ( ) always

6. How often do you have the nagging feeling that the intended audiences for your employee news or general employee information/communications programs may *not* be seeing them?
   a. ( ) never
   b. ( ) once in a while
   c. ( ) always

Okay, let's interpret your test scores. If the answer to Question 1 is either *b.*, or *c.*, and your remaining answers are mostly in the *a.* and *b.* columns, you are operating your network like a real pro. Now that may seem like a strange evaluation, considering the answers for Question 5, but we'll get to that later.

However, if your network is more than a couple of years old (*b.* and *c.*) and your answer are mostly in the *c.* and the *d.* columns ... you are suffering from "Network Narcosis".

What is "Network Narcosis"?

[*Nar.co.sis: act of benumbing; a state of stupor, unconsciousness or arrested activity.*]

It is a condition we are finding as more and more corporate video networks reach maturity. Its primary cause is overwork and

Distribution

complacency on the part of the video network manager. As networks mature, quite often their programming demands are greater. New clients and topics seem to be crawling out from under every organizational rock — Legal, Safety, Customer Service, Marketing, Accounting, Human Resources, Line Operations and so on. The momentum just keeps building. Yet, times being what they are, these programs are being produced with little if any additional staffing.

As programs proliferate, more and more names keep getting added to the program distribution lists, in many cases whether or not they really need or even want the programs. Videocassette duplication has become another version of the photocopying machine. Before the advent of that electronic wonder, the number of copies of letters and reports were limited by the number of carbons a typist could make. You thought hard about who really needed the information before adding a name to a distribution list. Now we photocopy everything and send it to everyone believing, in our hearts, that we are good, modern communicators.

The same thing can happen to videotapes, only the expense of wasted copies is a lot higher than with printed pages. In the rush of production it is easier to cover everybody on the list rather than to determine in advance just who should be getting what particular program.

Now, if no one in the field is complaining about the programming — its quality or content, the delivery of casssettes, their audience's reactions and so forth — you figure all must be well and good. You can take a breather.

Yet there is still that sensation that maybe things are not what they appear to be. Maybe the intended audiences for your programs are not really seeing them. If they are not, you have lost touch with the main element in your video operation. Maybe it is because you really have no control over your delivery system.

Now let's review the quiz questions and see how you can correct the condition or at least prevent it from taking hold.

Question 1 — When your network has been operating for three or four years you enter a period of maturation. Or, your network is approaching its teens and your second generation of playback equipment is now a mixture of U-matics, VHS and Beta equipment, all at least a couple of years old. In either case, you probably have not taken a good look at your network and how it operates since the serial numbers were recorded and the units shipped to the field.

In addition, if your network has grown by local offices buying their own playback equipment — as well as cameras for their own local program production — you probably do not know the real extent of your network. What to you, on paper, appears to be a 60-unit network may, in fact be a 100-unit network. In some cases extra copies of your programs are being made in field locations for additional distribution that will not show up on your records.

Yet you become complacent since you seldom hear complaints. The theory is "if it ain't broke, don't fix it". That's a sure symptom of "Network Narcosis".

Question 2 — If you update your field coordinators mailing list even once a year you will find that people retire, quit, get promoted, change departments and so forth. In some cases, whole divisions or offices get moved, sold, absorbed or eliminated.

Yet your mailing list, if it has not been checked within the last 18 months, contains these obsolete names and addresses. Worst of all, tapes may not be returned to you as undeliverable nor will you be notified of many of these changes.

Doing what the mail order houses call a list "purge" is not an easy thing to do. It takes time and some effort to organize, yet it must be done regularly. Mail order folks do it every 12 to 18 months. This is the only way to keep your list "clean" and to insure you that your tapes are actually getting into the hands of the people who are supposed to show them.

Send out a questionnaire to verify the names and mailing addresses of each person who is to receive a tape at least once or twice a year. If you get too many answers marked "retired" or "moved, left no forwarding address", then you know "Network Narcosis" has attacked. And where are the tapes you keep sending to these people? They're probably being used to record "Sunday Night at the Movies".

Questions 3 to 6 — Now, if we can assume that your tapes are at least getting to the right people, can we then assume that they are being shown to the people who are supposed to view them? NO! If you make that assumption, you have a very bad case of "Network Narcosis".

Tapes are being swept up in a corporate "Black Hole". Programs are received in some locations, dropped into someone's drawer and are never seen again. Eventually this can take you and your entire department into oblivion.

Distribution

Therefore, *make no assumptions about who views your video tapes*! Especially if you have not been in the field for over a year to visit network locations on fact-finding, not production trips. And especially if you keep getting responses like "your programs are just fine" on your feedback forms or in quicky telephone surveys.

No one likes your programs "just fine" all of the time. Sometimes they may really hate them or find them totally inappropriate for the intended audience at their viewing location. Remember, no two viewing locations are quite the same. And field viewing locations certainly aren't the same as those at headquarters.

Network managers have to get out into the field and meet with their coordinators personally — on their own turf — more than they do now. We know budgets are tight, but to keep an effective system running it must be constantly fine-tuned. Plant machinery has to be maintained . . . so does communications machinery.

A periodic check with the network coordinators should include:
— how tapes are handled once they reach the location;
— how they are shown (i.e., in meetings, on request, etc.);
— who sees them (i.e., managers, hourly workers, all employees);
— how they are promoted (bulletin board notices, memos, etc.);
— how they are scheduled (more than once, during lunchtime, etc.);
— where they are libraried; and,
— if and when they are recycled.

Also, coordinators should be asked about viewers' reactions to specific tapes (i.e., has the last tape on the reorganization or the latest acquisition been informative, or has it been a blow to worker morale?). Don't settle for "just fine" remarks. Too often in an attempt to be polite and to score points with headquarters, the field coordinators will not tell you that some of your programs have bombed . . . or, that the topic was too broad from the hourly employee's viewpoint . . . or, that morale is so low that the videotapes are scuttled by the plant manager in an effort to not show workers how you are wasting the company's resources and money. Talk with some of the managers and workers for whom tapes have been intended if you aren't satisfied with the answers you are getting from the coordinator.

Even if you have an established reporting system which evaluates each program that reached the field, you still need someone to get out into the field to do periodic checks. Even the best feedback systems have flaws.

Also, check to see if enough copies of a program are sent to cover all of the viewing facilities in each location when needed. You may be sending too few because they have equipment that's not on the corporate rolls or too many because many managers have their names on your list and not all of them really need to see each of the tapes.

Or, you many find that the managers at various locations are on "communications overload". Headquarters, divisional, regional and local training and communications programs all come in on videotape. We have seen instances of two tapes arriving on someone's desk, both addressing the same topic from two different viewpoints. In cases like these, no program gets viewed. Credenzas behind many a manager's desk are stacked with tapes that some day he hopes he will have a chance to review.

Finally, find out about the communications environment into which you are sending your programs. Is it appropriate for videotape viewing? If you are sending news or general information programs into a plant where they only have a half-hour lunch break, you may have a serious problem. Most employees want that time for themselves and resent the intrusion of company messages. In one plant we visited, the playback unit was turned to the wall.

Another problem is reaching employees such as salesmen, service personnel and other field workers who work away from organizational offices. Many only come into the office or plant once a week or even once a month. How do they see your program? Can they take a videocassette home to review? Ironically, since they are meeting people outside the company more than other employees, their information needs about your organization are much greater.

Generalizations cannot be made about all of your viewing locations. Information needs of the employees in each of your locations may differ widely. These needs must be understood and exceptions made when necessary.

In addition, you must find out about how much local management's autonomy plays a role in whether or not your programs get viewed by more than just a few selected managers who then judge whether or not their employees really need to see them. If the program is deemed important at the senders' level it can sometime be shelved at an office or plant because the local managers don't understand the complexity of the issue due to a parochial viewpoint.

Large networks are not the only one's who may fall victim to

"Network Narcosis." Small networks are just as susceptible to the malady. The problems are the same although their scope may be slightly different. Follow our guidelines and "Network Narcosis" will be easily arrested at the outset.

With this in mind, let us now take a look at the study findings on program distribution.

Viewing Locations

Only eight percent of the respondents today have 10 or less viewing locations for their programs (Table 10-1). In 1981, that figure was 26 percent while in 1973 it was 54 percent.

The median number of locations today is 70, more than double the number, 34, reported in 1981. Today's number radically differs from the early years of corporate video when the median number in 1973 was only eight (Table 10-2).

Once again, the most dramatic expansion in the number of video locations has taken place among the manufacturers. Since 1981, their median number of viewing locations has risen from 33 to 73, an increase of 121 percent.

Video Networks

Back in 1973 during the first study, we arbitrarily defined a video network as a distribution system which carries user-originated programming to six or more locations away from the point of distribution.

Table 10-1 informs us that nearly 97 percent of the video respondents have video networks in place. This figure has been steadily rising. In 1981, only 83 percent of those responding had networks, while in 1973 the figure was only 62 percent.

Those organizations with over 50 locations have increased from 38.8 percent in 1981 to 57.7 percent of the user respondents today (Table 10-3). In addition, we find that 38 percent of the respondents have over 100 viewing locations, compared to 22 percent in 1981.

Many of these large networks have been in place for a number of years, like Texas Instruments, 3M Co., Merrill Lynch & Co., Standard Oil Co. (Indiana), Prudential Insurance Co., GTE of Florida, Westinghouse Electric Corp., Allstate Insurance Co., and IBM Corp. Many of these companies have networks numbering well into the hundreds or thousands.

TABLE 10-1
Number of Viewing Locations In Which Private Television Programs Are Viewed

No. of Locations	Mfrs. N=78	Non-Ind. N=131	Med./Ed. N=7	Gov't./Other N=11	% of Total N=227
Less than 5	3.9%	1.5%	14.3%	9.1%	3.1%
5 to 10	3.9	3.8	28.5	18.2	5.3
11 to 20	11.5	6.9	14.3	18.2	9.3
21 to 35	15.4	18.3	14.3	9.1	16.7
36 to 50	6.4	8.4	—	18.2	7.9
51 to 100	19.2	22.1	14.3	—	19.8
101 to 200	15.4	12.2	—	—	12.3
201 to 300	7.7	9.2	—	—	7.9
301 to 400	3.9	3.8	—	9.1	4.0
401 to 500	—	2.3	—	—	1.3
501 to 1,000	2.5	4.6	14.3	—	4.0
Over 1,000	10.2	6.9	—	18.1	8.4
	100.0%	100.0%	100.0%	100.0%	100.0%

Distribution

TABLE 10-2

**Growth Rate of Viewing Locations—
Median Number 1973 vs. 1981 vs. 1985**

	Median Number of Viewing Locations			
	1973	1977	1981	1985
Manufacturers	7	15	33	73
Non-Industrials	14	21	39	74
Medical/Educational	—	—	8	15
Gov't./Other	3	14	26	28
Total Users	8	18	34	70

TABLE 10-3

**Summary of Number of Locations in Which
Private Television Programs Are Viewed 1981 vs. 1985**

	% of Total Respondents	
	1981	1985
No. of Viewing Locations	N = 325	N = 236
Less than 20	41.5%	17.4%
21 or more	58.5	82.6
51 or more	38.8	58.9
101 or more	22.2	39.0

Network Expansion

Almost half of the respondents, 49.4 percent, said that they planned to expand their video distribution network in 1986. The median expansion will be 10 percent, or an addition of about 8 more locations per network.

Eight seems to be the magic number in network expansion. In both the 1977 and the 1981 studies, the median number of locations in expansion plans was eight.

Selective Distribution

A phenomenon which we first noticed in 1981 and talked about in terms of "specialized networks" in the last book, has now fully evolved. While organizations may have numerous viewing locations available, only 42 percent of the survey respondents said that they send every program to every location. Fifty-eight percent responded that they send programs to a selective group. These may consist of supervisors, chemists, operations managers, sales people and so on who need to view specific programs.

The median number of locations to which the respondents send programs is 19. Just over half, 50.3 percent send their programs to 20 or less locations (Table 10-4).

TABLE 10-4

Average Number of Locations to Which Programs Are Distributed
N = 132

Average No. of Locations	% of Total
Less than 5	15.2%
5 to 10	18.9
11 to 35	28.0
36 to 100	15.2
101 to 500	18.9
Over 500	3.8
	100.0%

Thus organizations, hopefully, are using the medium more cost effectively by not scattering videotapes throughout the countryside as they would photocopies of seldom-read memos.

International Distribution

Program distribution to overseas locations is increasing. Almost one-third, 31.2 percent, of the respondents distribute their programming internationally. Of these, 73.4 percent distribute to 30 or fewer overseas locations and 9.1 percent distribute to over 100 foreign locations. In comparing these figures with the 1981 study, 80.5 percent had 30 or less locations internationally and only 4.8 percent fell into the 100 and over category.

The median number of locations in the respondents' international networks is 10, 37 percent higher than the 7.3 locations reported in 1981.

Therefore, international networking has grown considerably during the past four years and probably will continue to grow as organizational video becomes even more commonplace abroad, basically through the growth of consumer video.

Distribution

Languages In Which Programs Are Distributed

Seventy-six percent of the respondents distribute their programs in only one language, English, while another 13 percent use two languages. In addition, 7 percent of the respondents distribute in three or four languages and another 4 percent use more than four languages for their programs.

This does not mean they are distributing all of their foreign language programming overseas. Some organizations are using Spanish-language programming to reach their Hispanic workforce in the United States and sending programs in French to the Canadian Province of Quebec.

How Private Television Programs Are Viewed

If you have followed the tips outlined in the self-help quiz at the beginning of this chapter, you have a good grasp of the following section. You not only know the number of times your programs are viewed, but also *how* they are viewed.

As we have mentioned frequently, once your program is in the field — no matter how good it is, no matter how important it is — it may never be seen because of the viewing environment into which it has been sent. Local autonomy, workplace conditions, allocation of viewing time, shift work, and so forth can play havoc with your best laid plans.

The only sure way to know whether or not the programs are viewed and how is to do a field survey at least every two years, or if your organization is in a great state of flux, even more often.

In Table 10-5, we find that the percentage of program producers who don't know how their programs are viewed has decreased 6 percent, to 12.9 percent, since 1981. A few more are fighting *network narcosis*.

Small group sessions are the most popular form of program viewing in private television with more than two-thirds of the respondents reporting this method. This is about the same level as in 1981.

Two viewing methods experienced a notable change in the last four years. Viewing programs at the individual's discretion has lost some of its past popularity and now is reported by only 34 percent of the respondents, a decrease of 13 percent from 1981.

Viewing of programs by outside groups has increased some 20 percent over 1981. More organizations are using video to reach community organizations, government groups and other external audiences today than ever before. This is due in large measure to the penetration of consumer video units into offices and meeting rooms which, prior to the early 1980s, had no means of viewing videotapes.

Number of Times Programs Are Played

Over half of the respondents (57.8 percent) say that their programs play 4 or more times. This is up 6 percent over the responses to the 1981 study.

Organizations which play their programs one to three times represent 40 percent of the respondents, an increase of 11 percent from the last report.

Even though it only represents three percent of the respondents, it is still unnerving to find that there is an entry for "don't know" in Table 10-5. Is this another symptom of creeping *network narcosis*?

TABLE 10-5

How Private Television Programs Are Viewed
N = 240

How Program Is Viewed	% of Total
Individually, at viewer's discretion	33.8%
Small group sessions (Less than 10 people)	67.1
Large group sessions (more than 10 people)	34.6
Customer/point-of-sale locations	15.0
Outside group meetings	10.4
Don't Know . . . send out to field to use as they see fit	12.9

(More than one answer may apply for each respondent.)

Means of Program Distribution

Hybrid distribution networks are more prevalent now than ever before as seen in Table 10-6. Few respondents reported using only one means for program distribution, more often than not two or three formats were listed.

TABLE 10-6

Means of Private Television Program Distribution
(In Ranking Order)
N = 243

Means of Distribution	% of Total
1/2-inch VHS videocassettes	92.3%
3/4-inch videocassettes	73.7
1/2-inch Beta videocassettes	33.7
In-house CCTV and MATV	11.5
Videodisc	4.9
Satellite	4.1
Other (inc. film)	2.9
Microwave	1.2

(More than one answer may apply for each respondent.)

When asked about the possible problems of multi-format distribution, over two-thirds said distribution in more than one format was no problem. The 31 percent who claimed this was a problem cited inventory and cost control as the reasons for their objection.

The 3/4-inch videocassette as the *primary* distribution method has fallen to a level of just 32 percent from 1981's high of 53 percent, a decrease of 40 percent. A look at the data in Table 10-7 shows that the VHS format now dominates with 92.3 percent of the respondents using it for some or all of their networks. Some 55 percent use it as their *primary* medium of distribution. Beta-only networks remain at the same level as in 1981 with 11.6 percent of the respondents reporting them in use.

TABLE 10-7

Primary Means of Program Distribution
N = 242

Primary Means of Distribution	% of Total
1/2-inch VHS videocassettes	54.6%
3/4-inch videocassettes	32.2
1/2-inch Beta videocassettes	11.6
Videodisc	.8
All other combined	.8
	100.0%

Nevertheless the 3/4-inch U-Matic cassette continues to be the "workhorse" in distribution as it has in production. Over 70 percent of the respondents reported that they are still using it for all or some of their locations as they did in 1981. In that study the impact of half-inch VHS and Beta playback units in private television was just being felt with only 19 percent reporting using VHS and 14 percent reporting Beta usage.

There are three interesting points to note about distribution from the current study. First, there has been a five-fold increase since 1981 in the use of in-house CCTV or MATV for program distribution as more buildings and complexes get "wired up". In 1981 only 1.9 percent used in-house cable for distribution, while today 11.5 percent are using it.

The second point is the inclusion of videodisc as a means of distribution as reported by 5 percent of the respondents. It is also listed as the *primary* means of distribution by slightly under one percent of the respondents. When the last study was conducted, the videodisc was just a glimmer in the future and few saw it as a distribution means when queried at that time. Obviously, it is still not a potent force as a distribution means but remains a unique training and informational tool.

The third point is the addition to the list of satellites as a means of distribution. Although it is used by only 4 percent of the respondents as *one* of the methods of distribution, it is a sign of times to come for private television.

Distribution Equipment Investment

The investment in program distribution systems has changed significantly since 1981. With the continued penetration of the organizational marketplace by home video and its less expensive playback units, the organizations which do not have viewing equipment have dropped from 8 percent in the 1981 study to 3 percent today.

The median investment in playback equipment has risen almost 20 percent in the interim. In 1981 it was $46,500, while this study's median is $55,500.

Although two-thirds of the respondents have less than $100,000 invested in playback equipment — unchanged from 1981 — there is a shift in the $50,000 to $100,000 category, with 5 percent more having that amount invested this time (Table 10-8).

TABLE 10-8
Total Investment In Distribution/Playback Equipment
N = 241

Investment	Mfrs. N=82	Non-Ind. N=137	Med./Ed. N=8	Gov't. N=9	Other N=5	% of Total N=241
Under $10,000	17.1%	12.4%	12.5%	33.4%	60.0%	15.8%
$10,000-$25,000	21.9	13.1	25.0	11.1	—	16.2
$25,001-$50,000	15.9	13.1	37.5	22.2	40.0	15.8
$50,001-$100,000	15.9	20.4	25.0	11.1	—	18.2
$100,001-$200,000	21.9	19.0	—	22.2	—	19.1
$200,001-$500,000	4.9	11.0	—	—	—	7.9
$500,001-$1 Million	—	6.6	—	—	—	3.7
Over $1 Million	2.4	4.4	—	—	—	3.3
	100.0%	100.0%	100.0%	100.0%	100.0%	100.0%

Once again, the non-industrials have a larger investment in playback equipment than does the manufacturing sector. Twenty-two percent of the non-industrials reported investments of over $250,000, while only 7.3 percent of the manufacturers have that commitment.

Significant in the analysis is the fact that three-out-of-four of the respondents who have been involved in video for six or more years have investments of more than $100,000 in playback equipment. These are more mature video users with larger networks in place which may consist of older, more expensive "industrial" playback equipment (e.g., 3/4-inch cassette players).

Reuse of Videotapes

Sixty-eight percent of the respondents recycle their videocassettes an average of six times. This is drop of some 15 percent from the reported 80 percent of users recycling tapes in 1981. However, the number of times a tape is recycled has gone from four in the last study to six times today.

There were some video users who claimed they recycle tapes "until they fall apart", in some cases as many as 50 times. Those are extreme cases, as can be seen by the more conservative average calculated.

The drop in recycling is due to the highly competitive pricing of both the tape manufacturers and dealers in the private television and consumer markets. You can go into any K-Mart or grocery chain store and pick up a VHS videocassette for about $5.00. They have become disposable items for many people when the shelf-life of the tape is somewhat limited. It may cost more in telephone calls, shipping and inventory control to keep track of the videotapes than to consider them as disposable items.

Chapter 11
Promotion, Evaluation and Feedback

Program Promotion

These days even home video gets the hype and advertising dollars which were once reserved only for large-budget broadcast specials or series. Posters, reviews, display and "tune-in" style advertising fill the daily newspapers and general readership and business magazines. Even "How-To" and self-improvement tapes are touted like network specials. Everybody wants their "share" of the viewers' ratings.

The necessity of "promoting" audience viewing of private television programs is finally being recognized as an important part of the overall management of the video function. In the past studies, we found that program promotion was often overlooked. Audiences were largely captive since most of the programs were of a training nature.

With the steady increase in the number of communications programs, producers are finding that the audience is no longer guaranteed. Unlike most training programs, many communications programs deal with "nice to know" information as opposed to "need to know" information and viewing is frequently left to the discretion of the viewer. Thus, the producers have realized that they must adopt some of the techniques of broadcast television and motion pictures to get their viewers to even turn on the set.

In the first study, only a little more than one-half of the respondents did any form of program promotion. This figure jumped to two-thirds by 1977 and to 80 percent by 1981. Promotion has leveled off with 79 percent responding in the current survey that they used some form of program promotion.

Direct mail and memos is the most popular promotional form used with 56 percent of the respondents casting their vote in that category according to Table 11-1.

TABLE 11-1

Forms of Video Program Promotion Used to Stimulate Viewing

N = 182

Form Used	% of Total Using Promotion
Posters	28.6%
Fliers	32.4
Employee Publications	48.9
Bulletin Boards	28.7
Catalogues	33.0
Direct Mail/Memos	56.0
Telephone	20.3
Other (Previews, Schedules)	10.4

(More than one answer may apply for each respondent.)

Nearly half of the respondents use regularly scheduled employee publications as a means of publicizing their programs. A third of the respondents said they used catalogues and fliers to alert viewers.

Many respondents use more than one form of promotional media for each program. Sometimes a memo or direct mail piece may precede the program while a flier or poster may announce the date and time of a meeting to show the program. Or, a feature story may appear in an employee publication promoting the importance of the program, followed by a bulletin board announcement and a direct mail or telephone campaign to assure viewer turnout.

Promotion is especially important if discretionary viewing is involved. However, most producers find it just as much of a necessity with required viewing as well to emphasize the importance of the program and/or promote the receptivity of the message to the viewers.

Validation and Measurement of Audience Reaction

There's good news and bad news in the reports on private television users' attempts to validate program results and measure audience reaction.

Let's begin with the bad news. Only 68 percent of the respondents make any attempt to measure or validate their program results. This is a decrease of 11 percent compared with the 1981 respondents.

Now for the good news. More organizations are trying to validate the results or measure audience reaction to communications programs. Seventy-three percent of the respondents are once and for all putting aside the old axiom that the effectiveness of communications programs cannot be measured (Table 11-2).

TABLE 11-2

Private Television Users Validating/ Measuring Program Results

N = 162

Validating Results of:	% of Total Validating
Training Programs Only	23.5%
Communications Programs Only	17.3
All Programs	55.5
Some/Few Programs Only	3.7
	100.0%

All programs, both communications and training, are measured for their effectiveness by 56 percent of the respondents, while 17.3 percent say they measure communications programs only. In 1981, only 60 percent of the respondents measured all programs, while 9.6 percent measured only communications programs.

Measurements Used

The principal means of measuring viewer reactions to programs remains the questionnaire or feedback form. Eighty-two percent of the respondents use this means for testing their results. In 1981, 76 percent of the respondents used questionnaires.

The use of question and answer sessions for validation has dropped from 24 percent in 1981 to just 14 percent today. (Table 11-3)

Also of less importance in the current survey was pre-test/post-test measurements of viewer knowledge. This method of validation garnered just 19 percent of the votes this time, while in 1981 it was third ranked with 31 percent.

Big gains were seen in the use of random interviews as a means of validation in 1985 with just over a third of the users employing this method. In 1981 only 26 percent used random interviews.

New to the list of validation measures in this survey is data from interactive systems. Although little used at this point, since only 2.5 percent responded in this category, it will be a significant addition to validation systems in the future.

TABLE 11-3
How Audience Reaction Is Validated/Measured
In Ranking Order of Use
N = 162

Form used	% of Those Validating
Questionnaires/Feedback Forms	82.1%
Informal Comments	51.2
Random Interviews	34.0
Pre-Test/Post-Test of Viewer Knowledge	19.1
Phone Surveys	18.5
Question and Answer Sessions After the Program is Viewed	14.2
Viewing Logs	9.3
Other	4.3
Interactive Systems' Data	2.5

(More than one answer may apply for each respondent.)

Chapter 12
Budgets, Charge Backs and Cost Analysis

Size of Video Budgets

Video operating budgets, which include salaries, facilities and overhead, tape stock, duplication and distribution, equipment amortization and maintenance, outside services, purchased programs and so on, have experienced a considerable increase since the 1981 study.

Today's median budget is $224,000, a 66 percent increase over the 1981 median of $135,000. In 1985, 53.5 percent of the respondents had budgets of more than $200,000. In the last survey, only 35.1 percent had the same size budgets (Table 12-1).

TABLE 12-1

1985 Operating Budgets for Private Television
(Including Salaries and Expenses of Staff, Equipment and Facilities, Production Expenses, Outside Services, Purchased Programs, etc.)
N = 239

Operating Budget	% of Total
Under $25,000	2.9%
$25,001 — $50,000	10.5
$50,001 — $100,000	17.2
$100,001 — $200,000	15.9
$200,001 — $300,000	14.2
$300,001 — $400,000	10.1
$400,001 — $500,000	5.4
$500,001 — $1 Million	11.7
Over $1 Million	12.1
	100.0%

A comparison with video budgets of 1977 shows that today's budget is more than three times higher than the $72,000 reported at that time. Inflation played a large part in the increase. However after factoring in the decline in prices of equipment, the gains are still quite impressive.

Most significant are the changes in the upper and lower ranges of the operating budgets. Those organizations whose budgets are under $25,000 make up only 2.9 percent of today's respondents. In 1981, respondents in that category were 12.7 percent. While the trend continues with more smaller organizations using video, they seem to command larger budgets for their operations than did their predecessors.

At the other end of the spectrum, we find that in 1985, 39.3 percent of the respondents spent over $300,000. In 1981, only 23 percent spent that much, an increase of 71 percent.

A closer examination of years of video experience and size of operating budgets, shows that three-out-of-four of those whose budgets are over $300,000 have six or more years longevity in the field.

Budget Expenditures

For the first time, the survey asked the respondents to give a percentage breakdown of what expenditures went into their operating budgets.

Salaries take the biggest bite. Although the average share of the budget taken up by salaries is 43 percent over all, many respondents — especially utilities — report that 55 to 75 percent of their budgets go into salaries. One wonders how much their program quality, distribution and promotion suffer when over half of the budget goes into the payroll.

Outside services, including freelance and production services, take up 21 percent of the average budget. Another 15 percent goes into facilities and overhead, while equipment amortization eats up an average of 14 percent. Duplication and distribution make up only 10 percent of the budgets. Approximately seven percent each is spent on tape stock, maintenance and purchased programming.

Cost Per Program

Maybe we are coming to grip with reality as we grow older in this business. Seventy-nine percent of the respondents reported that their average program cost is over $2,000. That doesn't sound like much,

Budgets, Charge Backs and Cost Analysis

but it certainly is better than the reported averages in the last three studies. In 1973, 50 percent of the users claimed their average was under $1,000. It was obvious that most of them were only counting minimal out-of-pocket costs at that time.

By 1977, less than half reported costs of under $1,000 and one-third said they spent over $2,000. On closer examination of their budgets and the numbers of programs produced, we discovered *they were spending three times more per program than they said*. We then were able to come up with an adjusted industry-wide average program cost of approximately $6,000.

In 1981, half the respondents reported an average program cost of over $3,000. But the majority also indicated that they were only counting out-of-pocket costs. A more realistic figure was calculated to be between $8,000 and $9,000 per program.

In the current study, the reported median cost per program is $6,300, with more than half of the respondents reporting average expenditures of over $5,000. (Table 12-2)

TABLE 12-2

1985 Reported Average Cost Per Program
N = 233

Reported Average Cost	% of Total
Under $2,000	20.6%
$2,001 — $5,000	24.0
$5,0001 — $10,000	20.2
$10,001 — $15,000	17.2
$15,001 — $20,000	8.1
$20,001 — $35,000	8.6
Over $35,000	1.3
	100.0%

When asked what is included in their average program cost, only 45 percent of the respondents said it included "all costs" — that is, staff salaries, overhead, out-of-pocket expenses, production costs, equipment, facilities, outside services and so on. The rest (55 percent), responded that their program costs included "only out-of-pocket" or the direct costs for each production.

When all factors are taken into consideration a closer approximation of a median cost per program is between $9,000 and $10,000. This is substantiated by dividing the median operating budget by median number of programs which gives us a figure of $8,960.

In interviews with many seasoned in-house video producers, they say that they regularly are spending $20,000 or more per production. As it has been in the past, the average cost per program is a very gray area.

New Equipment Purchases

This also is the first study in which the respondents were asked to place a dollar figure on the amount of new equipment they had planned to purchase during the current calendar year. Ninety-three percent said they would be making equipment purchases during 1985.

Nine-out-of-ten of those making purchases said they were going to spend under $200,000. Two-thirds of the respondents said they were going to spend less than $50,000 during the year (Table 12-3).

TABLE 12-3
1985 Expenditures For New Equipment
N = 229

Amount Expended	% of Total
Under $25,000	47.2%
$25,001 — $50,000	17.9
$50,001 — $100,000	14.8
$100,001 — $200,000	10.9
$200,001 — $300,000	3.5
$300,001 — $400,000	2.2
$400,001 — $500,000	.9
$500,000 — $1 Million	2.6
	100.0%

The median figure for equipment expenditures was $29,250. Forty-five percent indicated that this was an increase over what they had spent in 1984. The average increase was 55 percent.

Less than half, 44 percent, said that they would increase their equipment budgets in 1986. However, the average planned increase was 73 percent.

Playback equipment was the first item on many users' 1985 acquisitions lists, with 45 percent saying they were to purchase these units.

Second highest on the shopping list was editing equipment, to be bought by 43 percent.

Third ranked was audio equipment said to be purchased by 40 percent. Other categories with more than one-third of the respondents saying they were planning to purchase included: cameras, production videotape recorders, monitors and auxiliary equipment.

The largest dollar expenditures per respondent were for editing equipment, cameras, teleconferencing and playback equipment.

Charge-Back Systems

When the video operation is a staff function serving various "clients" throughout the organization, budgeting and cost accounting is generally handled in one of the following ways:

1. Full charge-back: All costs, including salaries, overhead, depreciation, maintenance, production costs, raw materials, etc. are charged back directly to the budget or department or component "client" for whom the program is produced.

2. Partial charge-back: Only direct, out-of-pocket and production expenses associated with the production and distribution of each individual program are charged back to the "client's" budget.

3. Duplication/distribution charge-back: Only the direct costs of duplicating programs and distributing them to the receiving locations are charged back to the "client".

4. Straight budget: The video operation is given a set budget for an estimated or predetermined number of productions and all related services. There is no internal charge back to other budgets.

Seventy-seven percent of the survey respondents charge back some or all of their costs to their "clients" according to Table 12-4. This is an increase of 57 percent over the 1981 study, when only 49 percent had any charge-back system in place.

Sixteen percent more organizations were charging back *only* out-of-pocket costs and materials in 1985 than in 1981. Full charge-back as a practice has decreased almost 20 percent and distribution/duplication charge-back has dropped 35 percent in the interim.

TABLE 12-4

**Private Television Users
Having Charge-Back Systems
N = 181**

Charge-Back System	% of Total Charging Back
Charge-back all costs	24.3%
Charge-back only materials, direct costs	69.1
Charge-back duplication/ distribution	6.6
	100.0%

Needless to say, charge-back systems of any kind are controversial within most organizations. Some video users claim that charge back has a "chilling effect" on the use of video and would never be adopted in their organization. This attitude no doubt accounts for some of the 23 percent of the survey's respondents who do not charge back any costs or expenses. In these organizations, an annual fixed budget is determined and the video department attempts to serve their internal clients within it.

Such a system naturally breeds inequities since the video budget becomes a part of the general overhead of the organization and some components benefit from its services more than others. This also can lead to very nasty political situations.

The most popular charge-back system is a compromise with the video resources — people and equipment — being provided out of corporate overhead and direct program costs and duplication/distribution costs being assessed to the client or user. Seventy-six percent of those charging back costs fall into this compromise position.

When expenses on a production job are charged back, they often are passed on to client departments piecemeal and sometimes there is no summary record kept of the actual total cost of the job. Thus, many charges to the in-house client are disguised or buried in that department's budget. As a result, few really know what the *actual* costs of projects are. Most in-house clients, therefore, have developed unrealistically low expectations regarding the cost of video productions.

This has led to another misunderstanding on the part of clients regarding video program costs. That is the assumption by many that video is still "cheap" and that film production is "very expensive". It

Budgets, Charge Backs and Cost Analysis 119

also has caused some to contend that they don't need "film quality" . . . that video is "good enough" for their program.

Several factors have led to these conclusions. One is the simple fact that video is often done primarily in-house where all costs are not fully calculated as we have seen earlier on in this chapter. Most corporate film production, on the other hand, is done by outside production houses. All costs show up on the film producer's invoices. Second, many in-house productions are done with less sophisticated equipment than would be used by an outside producer.

In addition, in-house video production people are generally paid less than outside professionals. Therefore if a film is done outside it appears to be more costly than an in-house video production. It may look better as well. Thus, the unjustified reputation in some companies is that video is "fast and cheap".

One independent producer, Thoms H. Fraser maintains an extensive computer analysis of film and video production costs which he offers as a service for both his own clients as well as to outside private television users and commercial producers. It is based on his years of experience producing and directing films and video programs for commercial broadcasting and for corporate communications. He contends that film and video production costs are virtually the same for comparable quality when *all* costs are honestly computed.

To solve the charge-back problem, a major corporate video user approached it in this way. A year was spent tracking all costs so that a database for a full-scale cost accounting system could be established. Then, at the beginning of the new fiscal year, all jobs were invoiced to the clients reflecting two levels of charges. One was the direct out-of-pocket costs which were normally billed to in-house clients and the other was an itemized list of all other unbilled costs which were being absorbed by the organization.

The system ran for two full years, after which all clients were to be billed the full amount for each production. The two-year adjustment period was deemed sufficient to enable most in-house users to adapt to the full charge-back system and gain a working knowledge of how they should budget their productions. In this case, the full charge-back system did not work out after the test run, and the system continued with partial charge-back. Nothing was lost, but the clients did learn the *real* costs of producing a videotape.

If a full charge-back system is in place, the clients should have the option of utilizing the in-house video services or going directly to outside suppliers. They should be able to compare costs on a realistic basis as well as be able to make decisions on factors other than cost alone.

Cost Analysis

It is only when the true cost of a media program is known that its effectiveness can be measured against its predetermined objectives.

A good cost analysis program will include a method of not only monitoring "out- of-pocket" expenses, but will factor in staff salaries and benefits, office rent, machine depreciation and so on, into the *real* cost of production. Where feasible, any revenue increase or cost-saving to the organization through the use of the program also would be included in the analysis.

Organizations which have been using video for a long time are more likely to have formalized cost analysis programs — no doubt contributing to their longevity. And, interestingly, we find that video operations under the administration of Corporate Communications are more likely to have a realistic grasp of all of their costs. Generally, they are used to dealing with larger budget figures which come under closer scrutiny by management.

A contributing factor to the lack of good cost analysis systems in many video operations in our experience is that there is a similar lack in other areas of the same organizations. That is, there is frequently a lack of stewardship of budget figures at every level of middle management and little demand from the top for measurable cost justifications of most budget items.

Therefore, when video managers go to their accounting departments for assistance in setting up accurate analysis programs which include all cost factors, they often receive little or no help. It is just not part of the corporate culture to keep track of costs per square foot of space or monthly utility bills or supplies by each operating unit.

There is no mystique in setting up an effective costs analysis system. There are two basic cost categories to consider:

- **Annual Fixed and Operating Costs**

- **Out-Of-Pocket Production Costs Per Program**

Fixed, Operating and Production Costs

Fixed costs include:

1. Total salaries and benefits of all full-time staff members;

2. Offices and work space (Total sq. ft. x cost/sq. ft.)

3. Utilities (heat, light, air conditioning, etc.)

4. Annual equipment depreciation expense; and,

5. Corporate overhead (if assessed).

Operating costs include:

1. Equipment maintenance and repair;

2. Production and maintenance supplies (consumables);

3. Office equipment and supplies (consumables);

4. Communications (telephones, data, fax, postage, messengers, etc.)

5. Staff professional development (publications, memberships, tuitions, etc.); and,

6. Miscellaneous department expenses including travel and entertainment.

How video equipment is being carried on the organization's books should be of prime concern to the video manager. In some instances it is expensed, that is, written off in the fiscal year it is bought. In others it is carried as office equipment and amortized on one kind of a depreciation schedule. Or, it is considered capital equipment and part of the building and therefore is put on another kind of depreciation schedule.

The kind of equipment it is also has a lot to do with it. A full-scale, broadcast quality studio with 1-inch VTRs and CMX editing systems would normally be capitalized and written off over several years. On the other hand, half-inch field playback units would be carried as office equipment and expensed when purchased.

Production costs include such items as:

1. Professional talent;

2. Equipment rental;

3. Sets and props;

4. Art, graphics, music and sound effects;

5. Creative services (script writers, directors, etc.);

6. Outside services (extra crew, special post-production services, duplication, etc.); and,

7. Travel and lodging.

Cost Per Project

A basic element of computing costs for each individual project is the *production hour*. There are approximately 1,600 production hours in most organizations for each person and piece of equipment. This is 80 percent of the total number of working hours in a year and has built into it downtime, vacations, illness and so on.

This means you can calculate the actual cost per production hour for each of the fixed and operating expenses by simply dividing the total annual cost of each item by 1,600. Therefore, for every $10,000 of annual equipment depreciation expense, for example, $6.26 must be charged against each production hour.

If the total of all of your annual fixed and operating costs is $250,000, then your basic production hour cost is $156.25 for your people and facilities. All other costs and expenses are on top of this.

This fixed cost assessment must be made whether the job is done on your premises or not, with or without the use of outside services. *With fixed costs, the meter is always running whether the facilities are in use or not.*

From this point on it is very simple to derive a realistic "rate card" . . . which you should do in any event, whether you charge back all costs, some or none. Estimating or totalling project budgets becomes easy since you multiply the hours involved by the "production hour" rates of the people facilities and then add the out-of-pocket production costs.

Project Estimating

Each project should have a cost estimate checklist showing both in-house costs and rates for comparable outside services. All participants in a project should have a thorough understanding of the cost elements, including those that may involve overtime or changes and additions after the project has been started. Clients are quick to forget all the alterations they make in a program after you give them the initial estimate.

The Bottom Line

What we have said in the last two books on the topic of cost analysis once again bears repeating. It is something that no astute manager ever should lose sight of.

"There is little the video manager can do when the company falls on hard times and there are wholesale cutbacks throughout the entire organziation, but he or she should at least try to be positioned so that video is not the first to go. Nor is there much that can be done in the face of gross bad management at the top, except, of course, to try to survive until the inevitable management change takes place. A sound cost justification program may be a life preserver in these stormy seas.

"Users were asked what they thought was the best way to keep management sold on video. The most commonly expressed answer was 'produce good programs', followed by 'keep management involved'. There can be no argument with the validity of these two points, as far as they go. But they don't go far enough to protect the dollars invested in video by most organizations. It takes only one program to bomb and put a sour look on top management's face.

"If you are fortunate enough to have prospered under a management "Godfather", what happens if he leaves the organization, or there is a general management changeover and the new people do not know you or know what you have done? A thorough set of cost justification figures is important, even if you are never asked for them. As long as you are producing good programs and keeping management involved you may never be asked. But the day you are, you had better be ready.

"Remember, too, that the request may not come from the top, but from the financial controller's office. The controller may not have seen many of your programs nor have been involved in any of them.

"He will look at your operation from a purely financial standpoint. As the head of the video operation of one of the nation's largest and best-managed corporations said, 'For all intents and purposes, we're working for the accounting department, not management.'

"More and more, top management is coming to realize that communications within an organization must, like every other aspect of the business, be a managed function. And they will need a manager to run this function. They can either hire from outside or promote from within.

"Whether you want to position yourself as a specialist/technician or as a manager/communicator, it is up to you. If you choose the management career path, then you must acquire and develop the necessary management skills and be prepared to be evaluated by the same standards applied to other managers in the organization."

Chapter 13
Organization and Staffing

Where Video Fits In The Organization

"The greatest mistake this company could make would be to have a corporate-wide video policy with all video applications centralized and controlled at corporate headquarters. The fastest way to kill anything worthwhile is to <u>institutionalize</u> it." — Division President of a large manufacturing firm.

"Our current management philosophy is to push decision-making as far down into the organization as we can. This includes responsibility for communications as well." — Chief Executive of a major corporation.

Where to locate the video function in an organization is a question we are being asked quite often these days. As we have often reported, the trend over the last ten years has been to centralize the video function in the Corporate Communications Department. This is the umbrella for many of the communications and "relations" functions within most organizations.

Today, "Corporate Communications" embraces *Public Relations, Marketing Communications, Government Relations, Employee Information, Investor Relations, Community Relations* and often *Graphic Services* and *Audiovisual* and *Video* as well.

The rationale is simple. Corporate Communications is — or should be — the information crossroads of the organization. Here more of the internal and external information is prepared for dissemination to audiences both inside and outside the organization. It is also the area in which trained communicators are located, or to which they report.

This has come about because top management has realized that trained professionals should be dealing with the messages and media if the organization's story is to be gotten across properly to an increasing variety of audiences. Since top management is very interested in video and since Corporate Communications is more directly involved with top management on a day-to-day basis, it is natural that this has been the department into which the video function has fitted best so far.

Who's In Charge

In the early years of private television, video was under the direct administration of the Personnel and Training functions within many organizations. In both the 1973 and 1977 studies, 40 percent of the video operations reported to Personnel and Training Departments. By 1981, only one-third of the video people reported to Personnel. Today, this figure has dropped even lower to 22 percent. (Table 13-1)

At the same time, the four studies have chronicled the movement of video into the Corporate Communications Department. In 1973 only 20 percent of the video functions reported to Corporate Communications. By 1977 the figure jumped to 32 percent and by 1981, to 37 percent. Now 45 percent of the respondents report to Corporate Communications.

Forty-nine percent of the non-industrial respondents are administered by Corporate Communications, an increase of 16 percent from 1981, while 44 percent of the manufacturers' video units report to Corporate Communications, an increase of 19 percent.

The number of private television units reporting to Sales/Marketing/Advertising Departments has jumped slightly from 13 percent in 1981 to 16 percent today. Twenty-three percent of the video operations in the manufacturing segment report to Sales/Marketing, while only about half that number report to Personnel. Obviously, products are being introduced, demonstrated and sold using video.

An interesting and significant trend is spotted in the four percent of the respondents who reported that the video function is under Line Operations. This underscores the fact that more and more *end users* are taking control of the medium which we mentioned earlier.

In some cases, video has been acquired by Corporate Communications and included under its umbrella of functions when the Personnel Department put it up for adoption. Trainers realized that their primary job was to train people not to be a media production unit. In some

TABLE 13-1
Administration of Video Operation
N = 245

Video Operation Administered By	Mfrs N = 85	Non-Ind N = 138	Med/Ed Gov/Other N = 22	% of Total N = 245
Corporate Communications	43.5%	48.6%	27.3%	44.9%
Personnel/Training	13.0	26.8	31.8	22.5
Administration/Financial (CEO, President, etc.)	9.4	5.1	18.2	7.7
Marketing/Advt/Sales	23.5	13.7	4.5	16.3
Line Operations	5.9	3.6	—	4.1
Other (Nursing, Educational or General Services, etc.)	4.7	2.2	18.2	4.5
	100.0%	100.0%	100.0%	100.0%

cases, they gave up video and all audiovisual media productions to become "clients". Often this has led to bigger budgets for the video operation since Corporate Communications can generally command larger sums for their activities from top management. This in turn meant better equipment and higher quality video productions as well.

Other video units happily fell victim to internal corporate raiders who plucked the function out of Personnel and placed it in Corporate Communications. Many of these were power plays by corporate communicators who saw controlling communications as a way to build their own power bases.

Yet in still other organizations the video function came into being within the Corporate Communications Department as an employee communications tool. Its first task was to produce news and information programming. Once established, the medium was found useful for other areas of communications as well as for training. Although not the recommended way for video to get started, it has assisted in the medium's organizational entry in a number of cases.

Reporting Level

A significant increase in the level to which the video function reports is seen in the survey. Ninety-four percent of the respondents say that they report directly to a Department Manager or higher. In 1981 only 80 percent reported to the same level.

In addition, 55 percent of the respondents report to a Vice President or Division Manager or higher. The rise in the reporting level is indicative of the continuing interest in the medium by upper management and the recognition of its ability to achieve overall organizational goals and objectives. It is interesting to note that the non-industrial sector video units report to a higher level within the organization than do the manufacturers. Sixty-one percent of the non-industrials report to an Executive Vice President or Division Manager or higher while only 45 percent of the manufacturers report at a similar level.

Video Centralization

In two-thirds of the organizations responding to the survey, video is centralized in one department. Everyone else in the organization is their client.

More than one independent video operation exists in separate areas of the organization in another 20 percent of the respondents' situations.

Organization and Staffing 129

The remaining 14 percent of the respondents report that video production is dispersed throughout their organization and is carried out by the programs' end-users.

End Users As Producers

Should video be totally centralized . . . and in the Corporate Communications Department? Well, yes and no.

Obviously from the survey respondents' reports we see that video is not only reporting to such end users as "Line Operations" but in a number of cases centralized control of all productions is not a factor.

The early 1980s had two parallel corporate revolutions underway, both related to the end users and their control over certain functions. The entry of the desktop computer seriously impacted the end users' dependence upon the big mainframe computers. The introduction of low-cost, easy-to-use video cameras and recording equipment made it possible for end users to have control over the medium for their own sometimes unique applications.

With the proliferation of home VCRs, the video mystique has evaporated. Even in the smallest of towns you can find at least one or two outlets for rental of video equipment (sometimes even from the local drugstore!).

The effect it will have on centralized video is just beginning to be felt. What we are now finding in our field interviews is the attitude from communications and training managers: "We know what we need. Just leave us alone and we'll do it ourselves. We don't need help from Corporate headquarters."

As we pointed out in Chapter 8, there are many specialized video needs at the local level that do not require the expertise of trained video professionals, just as there are countless small word and data processing applications at the local level that do not require mainframes.

This does not mean the end for the video manager any more than it does for the DP manager. All they have to realize is that they are a staff service whose sole purpose is meeting the needs of the end users. If this means installing video cameras for the end user, then that is the name of the game.

Cooperation With Other Video Units

Cooperation with other units takes on various forms. All of the respondents in organizations where there is more than one video unit reported that they work with other video units within their organization. Fifty-six percent of these said that they share resources, ideas and equipment with other video units, while 44 percent said that they provide consulting and creative design services to other units.

Video Policy

The respondents are equally divided on the topic of a clearly defined policy on video usage. Half of the respondents said they had such a policy. Of those who indicated a policy was in effect, 83.6 percent said their policy encourages viewing of programs by their intended audiences while 77.1 percent said their policy requires that all video production go through one control point.

Table 13-2 also shows that 56.6 percent said their policy mandated uniform standards and formats throughout the organization. In only 38.5 percent of the respondents' organizations does the policy permit any department or division to produce its own program.

TABLE 13-2

Type of Video Policy In Effect
N = 122

Type of Video Policy	% of Respondents with Policy
Mandates organization-wide standards and formats	56.6%
Requires central control of video	77.1
Permits individual units to produce own video	38.5
Encourages or promotes video viewing	83.6

(More than one answer may apply for each respondent.)

Video Staffing

Video staff sizes have remained about the same over the last few years. The median size of a typical private television staff is now 3 people, down one-half of a person from 1981 when it was 3.5 people. (Table 13-3)

TABLE 13-3

Number of Employees Devoting All or Major Portion of Time To Video

N = 241

No. of Employees	% of Total
1 to 2 employees	34.9%
3 to 5 employees	39.4
6 to 10 employees	14.9
11 to 15 employees	5.4
16 to 20 employees	2.9
21 or more employees	2.5
	100.0%

Although one-quarter of the respondents have large staffs of 6 or more people, with a full complement of managers, producers, directors, writers, designers, engineers and technicians, most organizations maintain a rather lean video unit. The three person department is most likely to consist of a manager or possibly producer/media director with two of the following reporting to him or her: a production assistant, a writer, a technician and/or an engineer.

Thirteen percent of the respondents said that their staff size had decreased over the last two years, while 40 percent reported an increase. Forty-seven percent said their staff size had remained the same.

Job Titles

Job titles and descriptions vary widely from organization to organization. There are no universal standards. In Appendix B we have outlined a series of generic job descriptions which can provide some guidance in preparing descriptions for your organization.

Chapter 14
Interactive Video

Interactive video came upon the private television scene early in this decade with the introduction of microprocessor-based videocassette and videodisc controllers/indexers and low-cost computers and peripherals designed for both home and office use.

Unfortunately, its potential was clouded early on by excessive promotion, poorly designed programs, inappropriate applications and the eventual demise of the two leading videodisc manufacturers, DiscoVision (a joint venture of IBM and MCA) and RCA. QUBE, Warner Communication's interactive cable TV project also shut down around this time, further throwing the future of interactive video into doubt.

However several major companies such as Pioneer, Sony and 3M are currently making major commitments to this technology and their patience and persistence give hope for a bright future.

With the demise of DiscoVision, Pioneer Electronics bought the marketing rights and the U.S. replication facility and is continuing to promote both hardware and software. Sony markets a complete line of interactive videodisc players and is very active in advancing the interactive technology. 3M's optical recording activities include videodisc replication (offering both three-day and one-day turnaround times), compact disc (CD) replication and new applications such as "write once" (CD-ROM) and erasable optical media which enable users to create their own discs.

Interactive learning is not a recent concept. It began with Socrates' methods of questioning and continued with the programmed texts and "teaching machines" of the fifties. In the 1960s "Computer Assisted

Instruction" (CAI) was in vogue in many organizations and in the educational community. But the technology was often abandoned when organizations found that the complications of program design and production often made the interactive systems too cumbersome and too expensive. Sometimes the programs became obsolete before they were even used due to the long preparation time involved.

In addition, the computers used for early CAI were large centralized mainframe systems which required hard-wired terminals wherever instruction was to take place. Long before "distributed data processing" and desktop PCs, these CAI systems were heavily dependent on major capital equipment, another impediment to their widespread use.

As with Mark Twain, reports of the demise of interactive video were premature. Today, with the advent of inexpensive microprocessor chips, desktop computers, inexpensive videodisc players and portable eight-inch videodisc players, many computers are using interactive video — both disc and tape — for applications ranging from bank teller training to operator tours of nuclear plants to point-of-purchase consumer information on garden care.

There are transactional point-of-purchase terminals that let users check out rapidly at hotels, drop rented cars and order discounted catalog merchandise. Many of the programs have varying levels of interactivity, ranging from simple branching to highly sophisticated interactions between the user, the computer and the video program.

There are two major approaches to interactive video as it is being used today. The first is *branching*. The simplest form of branching (Level I) does not involve computer data on the videodisc. It uses chapter start and stop points to segment the content into sections or chapters, much like the discrete bands of music on an LP record.

Level II branching is accomplished by a computer program recorded on small portions of the videodisc's second audio channel. This program is loaded into a microprocessor built into the videodisc player when the disc is inserted into the machine. The user still responds to questions and choices on the video screen but far more sophisticated levels of branching are available including testing, scoring and multiple paths of complexity to compensate for learner differences.

In both levels of branching the individual views a short segment of pre-recorded video program and then is given a choice of two or more other segments to view by punching the appropriate numbers for the desired segment on a simple remote control keypad. The video

Interactive Video 135

player then shuttles to the selected segment and the program proceeds. This is being heavily used in technical and sales training applications from the largest and oldest network, General Motors, to the IBM Guided Learning Centers, to the Sizzler restaurant chain.

The second approach — Level III — involves computer-assisted interaction where the videodisc player becomes an extension of the CAI approach. It is the most sophisticated level of interactive program design and involves a video playback unit with an attached external computer or microprocessor which completely controls the program.

These Level III applications vary from simple PC computers to major micros and minis attached to the videodisc player. They often permit user responses from other than traditional keypads — touch screens, joy sticks, "mouse" controllers, keyboards or custom control responses such as inputs from an aircraft simulator.

The viewer's response on the keypad, touch screen or keyboard attachment automatically selects which segment the viewer will see next based on an analysis of the preceding responses. In addition, it is possible to keep track of the viewer's progress through the program, thus validating and measuring the results and even qualifying or disqualifying the viewer from progressing through the course.

In the simplest forms of branching, the viewer is in control of the selection and sequence of the program segments he or she watches. In the more sophisticated computer-driven form of interaction, the program can be designed to tailor itself instantly to the viewer's needs by the microprocessor. All information that is already known or is inappropriate to the viewer's needs is automatically eliminated.

There is no question that the appropriate use of an interactive medium is much more effective than the more traditional media often described by such terms as "linear", or "one-way" (or "boring" by some). However, as we found in our studies of how companies deal with interactive program design and production, it takes special talents in "authoring" that often are not found within the average video department.

As the term may imply, "authoring" in this case means the task of writing instructions for a computer to follow as it operates the video program in response to the user's input. Authoring is really composed of two parts: (1) *instructional authoring* and (2) *computer design authoring and coding*.

The "instructional design" component of authoring is not new to interactive video. Human Resource and Training departments have used professional instructional designers for years (sometimes in-house staff, sometimes vendor-provided). Several companies in addition to the disc player manufacturers offer authoring methods and systems that range from a simple flowchart template and instruction book to software and authoring tools that run on a wide variety of personal computers.

Just as producing effective video programs has always been a team effort of writer, producer, director, graphics personnel, camera operators, talent and so on, producing effective interactive video programs adds a complete authoring system to the team. In a good training situation, the instructional designer was already part of the team. Now he or she needs to have added competence in the design logic of interactivity.

Drs. Diane Gayeski and David Williams, principals of the interactive instructional design firm, OmniCom Associates, say in their recent book "Interactive Media" (Prentice-Hall, 1985) that "interactive video is a cross between computer-assisted instruction and traditional 'linear' video. As engaging as computer graphics and CAI can be, nothing beats the sound, motion and color video displayed on the TV screen."

They say interactive video is beginning to show its potential for becoming a highly effective adjunct to traditional teaching and self-instruction as well as an important research tool. They predict the production of a new generation of programs with the new technologies which can:

- *"simulate mechanical, organic or interpersonal processes allowing students access to additional practice in situations that would be impractical for them to encounter in actuality;*

- *"provide drill-and-practice and tutorial instruction incorporating audio, still and moving visuals and computer-generated text and graphics;*

- *"tailor themselves to a variety of levels of knowledge, skill or interest, branching to remedial or more advanced material or different examples depending upon a student's input;*

- *"incorporate existing film, video, slide, graphic, computer and/or text materials into one package, which by its design mandates active student attention and participation;*

- *"provide feedback to both the student and the managing instructor in terms of individual answers and overall progress;*

- *"open new avenues for behavioral research and psychological assessment through the instruction of less obtrusive measures, more vivid nonverbal stimuli, and adaptive, individualized testing.*

Interactive Video Is Growing

In the 1981 study, 41 percent of the respondents saw a need for interactivity for their programming. (One-half of that number wanted it on interactive videodisc.) The remainder of the respondents were divided between 28 percent who saw no need for interactivity and 31 percent replying that they did not know the capability of interactive disc or tape. The later was certainly understandable since confusion existed over various hardware systems at that time.

Today, 22 percent of the survey respondents report that they are now using interactive video. Another 34 percent said they are planning to use it in the near future. The remainder said they had no plans to use interactive video at all.

Thus, in a period of four years we have seen a number of companies progressing from "seeing a need to use" interactivity to actually using the technology to produce programming. Although the number is not large, considering the complexity and cost of interactive video, it does constitute a growing area within the private television industry.

Tape vs. Disc

The primary medium used by 30 percent of those who are using or will use interactive video is the videodisc, according to Table 14-1. Another 37 percent are using videotape as the medium for their interactive programs. Interesting to note is the combination of media reported in use or of future use. Twenty-two percent said they are either using or will use videodisc and videotape, videotape and videotex or videodisc and videotex.

When asked if the videodisc was a replacement for videotape in their organization, 39 percent of the survey respondents answered that it was for *some applications*. Nearly one-third said that the videodisc was used only for *new* programming and distribution applications. Twenty-nine percent of the video users said it was used for new applications as well as a replacement for videotape.

TABLE 14-1

Media Used For Interactive Video Applications
N = 120

Media Used	% of Total
Videodisc	39.2%
Videotape	36.7
Videodisc and videotape	10.8
Videodisc and videotex	10.0
Videotex	2.5
Videotape and videotex	.8
	100.0%

Numbers and Applications of Interactive Programs

The numbers of programs produced per user in the interactive mode in 1985 were few. Seventy-one percent said they produced three or less programs, while another 14 percent said they produced between four and six. Just four percent produced seven to 10 interactive programs and the remaining 11 percent produced over 10.

Sixty-two percent of the respondents reported that they would increase the number of interactive programs they produce in 1986. Thirty-three percent said they would produce the same number again this year. Only 5 percent claimed they were reducing the number produced in 1986.

Job and *skill training* are the top uses for interactive video with 71 percent of the respondents saying that they produced programs for these two applications. *Point-of-sale* and *marketing* applications rank second with 43 percent of the respondents producing programs in this area. Lowest ranked application is *general information programming* with only 12 percent of the users reporting its use for this purpose.

Locations Using Interactive Video

As we reported in 1981, interactive video programs — tape or disc — are not generally distributed widely *within* the organization. The failure to perceive this was one of the factors that ended up putting DiscoVision out of business. They based their entire business strategy on the concept that all private television programs are distributed widely throughout all organizations and that videodiscs would eliminate tape for this purpose.

Interactive Video 139

Virtually all of their market planning was based on endlessly researching the numbers and sizes of various corporate video networks and calculating the supposed cost savings of videodisc program distribution over that of videotape. Until the end they failed to perceive the true nature of the videodiscs as something altogether different from all other forms of video programming and not just a cheap way to send out programs.

Most of the customers who came to them to have interactive videodiscs produced only wanted a handful of copies. The company's pricing strategy was to do the mastering at or below cost and make it up on the quantity runs. Only there were not enough quantity runs to make the company profitable.

In the current study, seven-out-of-ten of the respondents said interactive video was used in *six or less location* in 1985, with 58 percent reporting its use in only one to three locations. Seventy percent of the interactive video producers made less than three copies of their programs.

A quarter of those reporting claimed over 10 locations. This underscores the fact that videodisc programs produced for *internal* use require only a few copies. It is only the program produced for *external* distribution (such as Ford's training programs for dealer mechanics) that require mass duplication.

Just over half (53 percent) of the respondents said the number of locations using interactive video would increase in 1986, while 47 percent said the number would remain the same. Only two percent planned a decrease.

Video Manager's Involvement In Interactive Video

Eight-out-of-ten of the survey respondents are involved in one way or another with the interactive programming being produced by their organizations. Only 19 percent said they are not in any way involved even though their organization was using interactive video.

Forty-seven percent of the video managers who reported some kind of involvement said they are responsible for all aspects of interactive video from program design through production and replication. On the other hand, 29 percent said they are only responsible for the actual production of the program. Another 17 percent said they were only responsible for program design and authoring or for production and design.

Few video managers (two percent) are responsible at present for *only* the programming or authoring functions. The departments which do the program design, programming and authoring of the interactive programs — when not done completely within the video unit — are Human Resources, Training/Instructional Design and Marketing/Sales. Some video managers also reported that "end user" departments are doing much of the work.

So far, the technology of interactive video has been more advanced than most organizations' ability to use it to its full potential. We now see this changing as more and more exciting new applications are introduced and people become adept at designing and producing the programs. The future now looks bright indeed.

Chapter 15
The New Technologies

For some time we have been talking about the coming together of all of the modern communications technologies in the workplace and have been urging video people to learn as much about them as they could. In particular we have emphasized the need to become familiar with and involved in computers and office automation.

Our advice seems to have been taken to some degree. While only 68 percent of the video respondents are using general purpose computers in their operations, 86 percent said they *are involved* in the planning of office automation in their organizations, particularly in the use of "Local Area Networks" (LANs).

This is good news. However we are concerned about the 14 percent who are *not involved* in their organization's office automation planning. These are the ones who show up at seminars and meetings and look blank when asked if they know what a "LAN" is.

Local Area Networks

Today there is little reason for the video professional to be uninformed about what is going on in the various communications technologies. Every general business publication, business sections of many major newspapers and a variety of trade publications have devoted extensive coverage to the new developments in telecommunications and office automation.

As we pointed out in an earlier chapter, a LAN is really nothing more than a wire or cable system that links computers and other office equipment. An in-house coaxial cable TV distribution system that connects a video playback unit to several remote monitors can also be considered a local area network.

This means that, *technically*, a coaxial cable linking computers is also capable of carrying a video signal. While technically possible, depending upon the type of LAN system used, it may not always be feasible, however. Furthermore, only 25 percent of the currently installed LANS use coaxial cable. The remainder consist of the common "twisted pair" wires used for most telephone connections. At the moment, this type of wiring system is *not* capable of carrying a video signal, although many companies are now working on ways to make this possible.

There are two basic types of coaxial cable local area networks: *baseband* and *broadband*. The former will carry only one signal at a time, the same as the cable connecting the *video output* of a VTR to a *monitor*. A broadband system is like regular cable television. That is, one wire can carry signals on a large number of different "channels".

The trend today is mixed with many users installing simple (non-video capable) twisted pair wires instead of coaxial cables which at least have the *potential* of carrying video signals in addition to voice and data.

Another form of LAN now emerging rapidly are the optical fiber systems which use tiny strands of glass fiber the thickness of a human hair. Information is encoded on a laser beam which is then transmitted through the fiber. Each of these glass fibers is capable of carrying more information than a broadband coaxial cable.

Unfortunately, the people responsible for specifying local area networks for their organizations rarely ever think of the potential video uses. This is why the video manager should take an active interest in his or her organization's office automation plans.

It may be difficult in some organizations for the video managers to break the barrier between video and telecommunications or data systems, but for their own survival and development it must be done, or the communications revolution will pass them by.

We asked the study respondents to give us their opinion about the desirability of a LAN system which would permit the delivery of full-motion color video along with data and other forms of communications to every terminal or workstation in their organization.

Only 11 percent of those answering said that such an installation was *not* desirable. The remaining 89 percent were more visionary. One-quarter said the installations would be *highly desirable*, more than likely envisioning the various types of communications needs they would fulfill with it. Another 22 percent said it would be *somewhat desirable*, while 42 percent said it was *possibly desirable*.

The key point in all this is that while the video people think the whole idea is really nifty, the systems people say they can not see any reason for wanting to do it, so they will not plan for it. Therefore, it is up to the video people to demonstrate the need.

Computers in the Video Department

The influence of the micro computer can be seen in many businesses today with the advent of "PCs" on many desktops. It certainly is a different picture from what we first saw in 1980, when some video managers had to sneak their Apple computers into work behind the backs of the data systems people and memos were being issued to halt the infiltration of these small "unauthorized alien units" in the workplace. At that time, systems people were feeling threatened by the end-user having control over his or her data processing.

Since private television and computers have been running on parallel tracks in most organizations for many years, it is natural to assume that as soon as the PC computer became available it was immediately adopted by most video operations. While 68 percent is encouraging, it still is not enough when all of the benefits of the current generation of desktops are taken into account.

The most common micro computers used in video operations are the IBM, Compaq, Zenith or a host of compatibles based on the PC-DOS or MS-DOS operating system. Sixty-one percent of the respondents said they are using single-user micro systems in their operation, while 15 percent are using multi-user systems. A third of the respondents said they were able to access their organization's central mainframe computer through local terminals and 17 percent reported using multiple-user, mini-computer systems.

Not surprisingly, the most popular current computer application among video users is word processing with 96 reporting this use. Slightly over half of the respondents are using computers for budgeting, estimating, cost control and record keeping functions.

Few other actual video management activities are now being performed by computer, however. Only 48 percent keep their equipment inventory on computer. Thirty-one percent do program element inventories (i.e., stock or file footage) and 25 percent keep their production schedules on computers.

When asked about what software they need to make their use of the computer more productive, video managers cited scriptwriting, budgeting, video inventory and project management as their top-most needs. There are many excellent computer programs now on the market that do these things, however it is apparent that too many video people are not keeping up with what is available.

Computer Graphics

In another section of the questionnaire we asked the video users about computer generated graphics. Over 60 percent of the respondents said they were using computer graphics in their departments. It is unclear, however, what proportion of these involve the use of dedicated systems as opposed to using general purpose computers employing graphics software. In any event, this in itself is encouraging, especially since *the remaining 40 percent said they would be using computer graphics within the next two years.*

In view of the current wave of cover stories and special issues of both video and computer trade magazines devoted to computer graphics this is not surprising. Video professionals are quick to see when some new technology will extend their creativity and their budgets.

Many seasoned video professionals we have talked with tell us that they see a great potential for computer-generated graphics and animation at all levels of program production. As one producer told us:

> *"We can now show on videotape things that it would be impossible to photograph, such as the inside of a jet engine in flight or a human heart pumping. The costs are dropping rapidly while the quality is rising steadily. This puts relatively professionally produced graphics and animation within the reach of most video producers.*
>
> *"I can see the day approaching when we will be producing most of our program footage on the computer and only inserting live footage occasionally to make a specific point . . . the same way we now insert graphics into live footage."*

The New Technologies 145

In another section of the questionnaire we asked how interested video users would be in a specially augmented PC computer that would perform a number of video production tasks in addition to running all of the standard software programs for word processing, spreadsheets, database management, bookkeeping and so on. The special video capabilities would include digitalizing the video input and providing a variety of special effects manipulations in addition to a virtually unlimited character generation and computer graphics capability. The system would also function as a post-production and editing control system.

Needless to say, about two-thirds of the respondents said they would be very interested in such a system without even knowing the cost.

While there are several systems in the development stage that will deliver this capability, and at a price within range of nearly any video operation, the reaction of the respondents demonstrates the strong interest in and need for affordable video graphics systems. The fact that these systems can perform many general purpose functions in addition to the specific video and graphics functions will have a strong appeal to those who sign the checks.

Chapter 16
Satellite Videoconferencing

After over 10 years of relatively little activity, videoconferencing is now experiencing a resurgence in interest and growth. We call videoconferencing a "born again" technology since our first involvement in what is now called private television was with videoconferencing in the late Sixties and early Seventies.

At that time, nationwide, big-screen closed-circuit videoconferences were in vogue. As we said in Chapter 1, this was just about the only form of video distribution there was until the 3/4-inch Sony U-Matic videocassette was introduced in 1972. The alternative was shipping around rolls of half-inch or one-inch videotape which usually required as much technology and technical expertise at the playback site as in the studio. It was never very popular.

It was much more popular and exciting to get together in a big hotel ballroom with several hundred other people and have someone who had a lot to say about how you were going to be earning your living peering at you live from a 20-foot screen.

The videoconference applications then were very similar to those being used today. We saw the Democratic party launch a major fundraising campaign, Brown-Forman Distillers introduce its Holiday promotion to dealers, Thomasville Furniture show dealers how to sell, automotive executives tell their dealers to "sell or else!", cardiologists explain the latest in by-pass procedures to colleagues around the country, government officials interpret new environmental laws to industrial leaders and many more dramatic, high impact presentations.

Not only was videoconferencing very exciting, but at that time it was also *very* expensive. Only big companies could afford it and they rarely did it more than once a year.

On the other hand, it also was *very effective.* It was a highly interactive event staged in real time with everyone involved getting the same message. The event and its staging produced a major impact on the viewer. *Something was always accomplished.*

But the videocassette completely changed everyone's perceptions of video. For the first time the viewer could control the medium. Convenience and distribution became more important than immediacy and the impact of the big screen. Videoconferencing went into a decline.

Rebirth came in the late Seventies and early Eighties as satellite capacity increased and it became far less expensive to send a television picture on a 45,000-mile round trip into space and back than 3,000 miles through coaxial cables owned by the telephone company.

At the same time new devices called "Codecs" (for "code" and "decode") came out of the laboratories that would take the signal for a full-motion, color television picture and squeeze it down to the size of a standard data transmission channel and move it from location to location the same way computer information was moved.

Systems In Use

There are now *two* forms of videoconferencing. The original, sometimes called "ad hoc," "one-way" or "point-to-multipoint" videoconferencing, and the new form called "dedicated" or "two-way" or "point-to-point" videoconferencing. The differences are major.

A one-way *ad hoc* videoconference is more like television as we know it. It is an event — a "program" — and many of the talents and skills involved in an ordinary video production are employed, including those of the in-house video manager/producer.

While pre-produced videotapes are sometimes used either as the program itself or as program inserts, most videoconferences are "live" productions, just like the early days of broadcast television. What goes out is whatever happens in front of the camera at that moment, mistakes and all. Most one-way videoconferences also use two-way audio links with each viewing location, thus permitting all members of the audience to have the chance to participate in any discussions.

Most one-way videoconferences employ a full bandwidth analog

Satellite Videoconferencing

video signal of the type used in regular television, which means that the only quality limitations are in the cameras and display systems that are used.

In the early days of satellite videoconferencing a very large and expensive receiving "dish" was required. Today many videoconferencing signals are transmitted through satellites employing a higher frequency known as the "Ku band" which may require a much smaller, less expensive dish antenna. This makes a larger number of fixed receiving locations more attractive to active one-way videoconference users. According to Elliot M. Gold, publisher of "Telespan", the foremost publication in the videoconferencing and teleconferencing industry, "the fastest growing area today is dedicated broadcast video networks (DBS)."

There are a number of firms specializing in setting up videoconferences for companies and other organizations. Two of the largest, VideoStar Connections, Inc. and Private Satellite Network (PSN), offer clients a full range of services from occasional, *ad hoc* videoconferences to permanently installed networks with fixed reception and display set ups and, in some cases, permanent origination systems ("uplinks"). They also offer reduced cost links with many overseas locations through recent agreements with several foreign telecommunications operators.

As equipment and satellite transmission costs decline, more and more organizations are finding that videoconferencing can offer many benefits. The trend in this area is for many organizations to hold videoconferences with audiences outside of their own employees, dealers or stockholders.

For example, Texas Instruments which has been actively using videoconferences for over 15 years, recently held a North American videoconference on the hot new topic of "Artificial Intelligence" which was carried on three different satellites and seen in more than 500 locations by over 30,000 viewers outside the company who were interested in the subject.

Two-way videoconferencing is more like an ordinary meeting with a couple thousand miles running down the center of the conference table. At least it tries to be like an ordinary meeting and this is where the technology can make or break the day. Like a meeting, it is not an "event" and therefore is a lot less structured than a one-way production.

The participants can talk and see and interrupt each other, just like at a real meeting. The only difference is that, depending on the system, you only can see the person speaking and not the reactions of the others listening in the same room with that person (vital in many corporate situations where you do not know what to think until you can see what others may be thinking). Or, if you *can* see everyone at the other end, the image may be too indistinct to be able to catch all the expressions and nuances.

Two-way videoconferencing set-ups are usually permanent, "dedicated" facilities costing about as much as a fully equipped video studio. They usually are designed and installed by systems and telecommunications engineers who have never seen the inside of a TV studio and who also know very little about interpersonal communications. Too often, the expertise of video professionals is not sought (nor even welcomed) in the design of these facilities. This, of course, is a harsh generalization for which there are a number of pleasant exceptions, usually where the video people have been involved.

Except for the cameras and monitors (or projectors), the rest of the "black boxes" involved in most two-way systems are part of an alien technology called *telecommunications*. The gentle flowing waves of analog video are "digitalized," that is, converted into the bits and bytes of the data stream. Our faces and our voices are shipped from point to point, intermingled with accounting records and inventory data.

Marvelously, we are reassembled, hundreds of miles away. But it is never quite the same since the little black boxes (Codecs) transmit only about 20 percent of the information in each video frame and only that information which changes from frame to frame. This is much less expensive than transmitting a full bandwidth video signal in both directions. Consequently, the "compressed" picture is never quite as good as what we see on regular TV. But for the convenience, and the economics involved, a slight loss in picture quality is well worth the price.

Numbers of Videoconferences and Locations

Thirteen percent of the respondents to the survey said they currently are using videoconferencing. Another 11 percent said that their organization has plans for using it soon, most within the next two years. Twelve percent also reported that their organization has used videoconferencing in the past, even though they were not now using the medium.

Satellite Videoconferencing 151

Of those respondents using videoconferencing, over one-half (53.8 percent) said that they held one-way *ad hoc* videoconferences, while 21.2 percent held two-way, point-to-point conferences. One-quarter of the organizations held both types of videoconferences.

The median number of one-way videoconferences held in the last two years was two per respondent. The median of two-way conferences was four. There are videoconferencing users in both categories that report high levels of usage, but they are few in number. Only 16 percent of those using one-way videoconferencing have held more than 10 in two years. Thirty-nine percent of those using two-way videoconferences have had over 10 in the same period.

The median number of locations to which an *ad hoc* videoconference is beamed is six. Although some respondents reported over 200 locations (these are the ones most heard about), those used for smaller, selective audiences are the norm.

In over half of the organizations reporting, the video department is responsible for the planning and production of a videoconference.

Primarily a Management Medium

Almost two-thirds of the respondents said that top management through first-line supervisors were the primary participants in one-way videoconferences. Primary use of these conferences was for dissemination of important organizational policy information to this level of audience. Other reported uses of one-way videoconferencing were marketing and sales meetings and educational and training sessions for supervisors and management.

In the area of two-way videoconferencing, ninety percent of the users reported that senior and middle management are involved in *two-way* videoconferences. The top ranked use for two-way videoconferences was for management review and planning sessions. The second most popular use was for management and employee information. Respondents also reported sales and marketing meetings, training programs, engineering and production planning sessions as key applications for two-way videoconferences.

Ninety-five percent of the respondents said that the audience was "somewhat" to "very satisfied" with the quality of communications provided in the one-way videoconference. Most of those dissatisfied gave black marks to the production quality of the program itself ("too long," "poorly prepared," "boring", etc.) rather than to the information contained in it or the technical quality of the transmission and display.

Highest votes were given for the *immediacy* of the videoconference. Everyone had an opportunity to get the same information at the same time in the same way.

All indications are that the entire field of videoconferencing — both permanent one-way network systems and dedicated two-way systems — are in for a period of explosive growth over the next several years. A critical need for this growth to occur, however, is for an industry-wide concept to market videoconferencing to all types of organizations.

Chapter 17
The Market

As we have said in the last two reports, it is not our purpose in these books to provide detailed market figures and breakdowns on various categories of video equipment and services. That information is available only to individual organizations through custom market studies.

Nevertheless, we believe it would be helpful to both video users and suppliers for us to attempt to describe the overall size and general makeup of the industry as we see it.

Each time we do this, we find it to be increasingly difficult, however. When we published our first study 13 years ago it was easy to identify the approximately 300 private television "user/producers." Everyone in the business knew everyone else. It was relatively simple to find out what was going on and who was doing what. At that time virtually everyone producing private television programs had their own equipment and/or studios and the dollar volume of the market was not very difficult to calculate.

Today that world has exploded to almost 30 times its original size. The old direct lines of information and communications between video users no longer exist as they once did. Where once corporate studios were easy to count, the increasing use of portable equipment for field production and the abundance of rental equipment and outside services available even in the smallest market make it increasingly difficult to even define what a "studio" is.

The blurring of the lines between consumer, broadcast and corporate (or "industrial") video equipment present obstacles in deriving an accurate picture of what is spent for private television and what is not.

New users entering the world of private television do so with little or no equipment. In addition, the older operations are using home video cameras and playback equipment in field locations in addition to their central studios.

This plays havoc with keeping track of equipment sales, which was once a help in assessing the market. In the past, video equipment from professional dealers that did not go into the broadcast and cable market was sold to video users in business and industry, government, education and medicine. There was little or no consumer video equipment to cloud the picture.

Another factor which has become more prevalent in recent years is the continued decentralization of video within many organizations. There may be two or more video operations in some organizations, each with its own production equipment and budget. Communicators or trainers in local offices can run out to the local discounter to pick up a playback unit, TV set and maybe a camera during a lunch break and bury it on their expense accounts.

Finally, a factor which impacts all dollar projections is the effect of inflation over the last 13 years in salaries, overhead and operating expenses as well as equipment purchases.

Consequently it is more difficult than ever to keep account of everything that is going on. The networking, which was one of the points of pride in the industry as it emerged, is not as efficient as it once was. New users are hard to find and the old ones often are buried in the waves of corporate mergers now taking place.

After 15 years of tracking the private television industry we have come to the conclusion that the only meaningful data on the market can come from an analysis of *video user* patterns and purchases, rather than import figures and/or sales to dealers. These are no longer reliable nor do they provide a realistic picture of this ever-growing industry.

Size Of The Industry

Once again we have recalculated our market figures going all the way back to 1973 and we found that they hold up under scrutiny.

While the industry continues to grow at a healthy rate, we are now seeing a maturity in that growth. In the early years with fewer users in the market, the growth was in excess of 30 percent annually and the

The Market 155

industry doubled every three years. We now see an overall growth rate of between 15 and 20 percent per year. While not as dramatic as it was 10 years ago, it is still a healthy growth figure.

With all of the above factors in mind, *we estimate that the total dollar volume of the industry at the end of 1985 was $3.8 billion.* This includes everything from salaries and services to equipment and tape stock. We estimate that approximately one-third of this total represents equipment and supplies. By the end of 1986, the industry volume should reach $4.5 billion.

While we have consistently resisted making projections beyond three years, we have extended our forecast this time to the year 1990, just to round out the decade. At that time we project the total industry volume to be approximately $7 billion a year.

APPENDIX A

Survey Questionnaire

PRIVATE TELEVISION COMMUNICATIONS CONFIDENTIAL QUESTIONNAIRE: 1985

This is the **fourth** industry study of how video is being used for communications and training by business and non-profit organizations. **All information is confidential and will be separated from your organizational identification in the database when processed. No organizations will be identified with any of the data in the published report.** If you are not directly responsible for a video operation in your organization, please direct this questionnaire to the person who is. Please return the questionnaire in the enclosed envelope and you will receive the results of the study at **one-half the pre-publication price.**

☐ Please send me the "Fourth Brush Report" with the results of the study.
( ) ITVA member: $ 9.95   ( ) non-member: $ 14.95

Organization Name: _____

Your Name & Title: _____

1. Primary business/organization classification:

   1 ( ) Manufacturing
   2 ( ) Natural Res.(petrl/mining/frst.prod.)
   3 ( ) Construction/Real Estate
   4 ( ) Communications/Electronics
   5 ( ) Insurance
   6 ( ) Banking/Other Financial
   7 ( ) Utility (telephone,elec.,gas)
   8 ( ) Transportation/Shipping
   9 ( ) Food/Beverage
   10 ( ) Wholesale/Retail
   11 ( ) Travel & Lodging
   12 ( ) Publishing/Entertainment
   13 ( ) Service Industry
   14 ( ) Conglomerate
   15 ( ) Medical Services
   16 ( ) Education
   17 ( ) Government
   18 ( ) Other _____

ORGANIZATION & ADMINISTRATION

2. Total number of employees in your organization ...
   1 ( ) less than 500   4 ( ) 5,001 - 10,000   7 ( ) 50,001 - 75,000
   2 ( ) 501 - 1,000   5 ( ) 10,001 - 25,000   8 ( ) 75,001 - 100,000
   3 ( ) 1,001 - 5,000   6 ( ) 25,001 - 50,000   9 ( ) over 100,000

3. How long has your organization been using video?
   1 ( ) less than 1 yr.   3 ( ) 2-4 yrs.   5 ( ) 6-10 yrs.
   2 ( ) 1-2 yrs.   4 ( ) 4-6 yrs.   6 ( ) over 10 yrs.

4. Is your video operation under the administration of ...
   1 ( ) Corp. Comm./PR   3 ( ) Admin./Financial   5 ( ) Line Operations
   2 ( ) Personnel/Training 4 ( ) Marketing/Sales   6 ( ) Other_____

5. What is the highest ranking person to whom your video function reports directly?
   1 ( ) CEO/Chr./Pres.   3 ( ) VP/Division Head   5 ( ) Section Supervisor
   2 ( ) Exec. or Sr. VP   4 ( ) Dept. Dir./Manager 6 ( ) Other_____

6. What was the highest level of management that made the **final** decision to use video initially?
   1 ( ) CEO/Chr./Pres.   3 ( ) VP/Division Head   5 ( ) Section Supervisor
   2 ( ) Exec. or Sr. VP   4 ( ) Dept. Dir./Manager 6 ( ) Other_____

7. Does the Chief Executive of your organization ever appear on your programs?
   1 ( ) Never   2 ( ) Sometimes   3 ( ) Often

8. Has an organization-wide needs analysis study been done to determine how video can be best used for communications and training?
   1 ( ) No   2 ( ) Yes, by internal staff   3 ( ) Yes, by outside consultant

9. Does your organization have a clearly defined policy on how video is to be used and who is responsible for its implementation?   ( ) Yes   ( ) No

10. If your organization has such a policy, does it ... (check all that apply)
    1 ( ) Mandate uniform standards and formats throughout the organization?
    2 ( ) Require all video production to go through one control point?
    3 ( ) Permit any department or division to produce its own programs?
    4 ( ) Encourage the viewing of programs by their intended audiences?
    5 ( ) Other points: _____?

11. In some organizations all video activity is centralized in one department. In others, there are two or more independent video operations in separate areas of the organization. In still others, production is dispersed throughout the organization and is essentially carried out by the end-users. Which most closely describes how video is handled in your organization?
    1 ( ) All video centralized in one department
    2 ( ) More than one video operation No. _____
        Under administration of: _____
    3 ( ) Users produce own programs
    4 ( ) Combination of ___ and ___ above.

12. Does your operation work with the other video units in your organization in any way, such as providing consulting services, sharing resources, etc.? If so, how?
    1 ( ) Do not work with other video units   2 ( ) Work with other units
    How? _____

13. Check the related media technologies that are used by your department in addition to video ... OR, that you will be working with in the next 2 yrs. if not now ... OR, that other departments are currently working with.

	Are Using	Will Use	Other Depts.		Are Using	Will Use	Other Depts.
Teleconferencing	( )	( )	( )	Multi-image	( )	( )	( )
Interactive videotape	( )	( )	( )	Film production	( )	( )	( )
Interactive videodisc	( )	( )	( )	Audio cassettes	( )	( )	( )
Cpter-based instr.	( )	( )	( )	Print/Graphics	( )	( )	( )
Computer graphics	( )	( )	( )	Displays	( )	( )	( )
Slides	( )	( )	( )	Other _____	( )	( )	( )

    ( ) None of the above. We use only video.

14. How many staff employees in your operation devote all or a major portion of their time to video and in what capacity? (Fill in no. next to position.)
    Managers ____ Producer/Media Dir. ____ Writers ____ Prod. Assistants ____
    Supervisors ____ Directors ____ Artist/Designers ____ Engineers ____ Technicians ____

15. Over the last two years, has the size of your video staff ...
    ( ) increased ( ) decreased ( ) remained the same

16. Has your organization used or is considering using outside consultants for any aspect of the planning or set-up of your video operation?
    1 ( ) Have used consultants   2 ( ) Considering using consultants
    3 ( ) Would not use consultants (Why not?) _____

17. If your organization has used or is considering using outside consultants, which of the following areas would be involved?
    1 ( ) Communications Needs Analysis   5 ( ) Equipment Specification/Facilities Design
    2 ( ) Training Needs Analysis   6 ( ) Interactive Program Design
    3 ( ) Video Organization & Staffing   7 ( ) Teleconferencing
    4 ( ) Program Planning & Evaluation   8 ( ) Other _____

18. If are you using general purpose computers in running your video operation, check the applications you are using.
    1 ( ) Word Processing   4 ( ) Equipment Inventory
    2 ( ) Budgeting/Estimating/Cost Control 5 ( ) Production Scheduling
    3 ( ) Program Element Inventory   6 ( ) General Record Keeping
            7 ( ) **Not** using computers for general operations

19. For these applications are you using ...
    1 ( ) Terminals connected to a central main frame?
    2 ( ) Multiple-user mini system?
    3 ( ) Multiple-user micro system?
    4 ( ) Single-user micro system?
    5 ( ) Other _____

20. What type of applications software program that you do not now have would greatly assist you in running your video operation? _____

Survey Questionnaire 159

BUDGETS & EXPENDITURES

21. In 1985, what will be the total **Operating Budget** for your video operation? This includes staff salaries and expenses, amortization of equipment and facilities, production expenses, outside services, freelance personnel, supplies, overhead, purchased programs, etc.
    - 1 ( ) under $25,000
    - 2 ( ) $25,000 - 50,000
    - 3 ( ) $50,001 - 100,000
    - 4 ( ) $100,001 - 200,000
    - 5 ( ) $200,001 - 300,000
    - 6 ( ) $300,001 - 400,000
    - 7 ( ) $400,001 - 500,000
    - 8 ( ) $500,001 - 999,999
    - 9 ( ) over $1 million

22. How much higher or lower is this than your 1984 Operating Budget? (+/-)____%

23. By what percent do you expect your 1986 Operating Budget to increase or decrease? (+/-)____%

24. Approximately what percent of your 1985 Operating Budget will be spent on the following:
    Salaries ___% Facility ovrhd ___% Tape Stock ___% Duplication/Distribution ___%
    Equip. Amort. ___% Maintenance ___% Outside Services ___% Purchased Progs. ___%
    Other ___% (pls specify)_____

25. In 1985, approximately how much will be spent **for the purchase of new equipment**?
    - 1 ( ) under $25,000
    - 2 ( ) $25,000 - 50,000
    - 3 ( ) $50,001 - 100,000
    - 4 ( ) $100,001 - 200,000
    - 5 ( ) $200,001 - 300,000
    - 6 ( ) $300,001 - 400,000
    - 7 ( ) $400,001 - 500,000
    - 8 ( ) $500,001 - 999,999
    - 9 ( ) over $1 million

26. How much higher or lower is this than what was spent in 1984? (+/-)____%

27. By what percent do you expect your 1986 equipment purchases to increase or decrease in relation to 1985? (+/-)____%

28. Approximately what percent of this year's equipment purchases will be spent on the following:
    Cameras ___% Graphics Cptrs ___% Plybak Equip. ___% Teleconferencing Equip. ___%
    Prod. VTRs ___% Audio Equip. ___% Monitors ___% Satellite Rec. ___%
    Edit Equip. ___% Lighting Equip. ___% Mobile Unit ___% Computers ___%
    Switchr/CG/SEG ___% Aux. Equip ___% Facil. Constr. ___% Other _____ ___%

29. What is your organization's present total investment in cameras and other in-house production equipment and facilities (other than used for editing and completion)?
    - 1 ( ) none in-house
    - 2 ( ) under $10,000
    - 3 ( ) $10,001 - 25,000
    - 4 ( ) $25,001 - 50,000
    - 5 ( ) $50,001 - 100,000
    - 6 ( ) $100,001 - 250,000
    - 7 ( ) $250,001 - 500,000
    - 8 ( ) $500,001 - $1 million
    - 9 ( ) over $1 million

30. What is your organization's total investment in in-house editing and post-production facilities and equipment?
    - 1 ( ) none in-house
    - 2 ( ) under $10,000
    - 3 ( ) $10,001 - 25,000
    - 4 ( ) $25,001 - 50,000
    - 5 ( ) $50,001 - 100,000
    - 6 ( ) $100,001 - 250,000
    - 7 ( ) $250,001 - 500,000
    - 8 ( ) $500,001 - $1 million
    - 9 ( ) over $1 million

31. What is your organization's total investment in playback equipment (VCRs, monitors, etc.)
    - 1 ( ) none in-house
    - 2 ( ) under $10,000
    - 3 ( ) $10,001 - 25,000
    - 4 ( ) $25,001 - 50,000
    - 5 ( ) $50,001 - 100,000
    - 6 ( ) $100,001 - 250,000
    - 7 ( ) $250,001 - 500,000
    - 8 ( ) $500,001 - $1 million
    - 9 ( ) over $1 million

PROGRAMMING

32. Is your organization using, or planning to use, video for any of the following program applications?

	Are Using	Will Use		Are Using	Will Use
a. Skill Training	( )	( )	j. Employee Orientation	( )	( )
b. Job Training	( )	( )	k. News Programs	( )	( )
c. Sales Training	( )	( )	l. Employee Information	( )	( )
d. Prof. Upgrading	( )	( )	m. Management Communications	( )	( )
e. Superv. Training	( )	( )	n. Economic Info/Education	( )	( )
f. Managment Develop.	( )	( )	o. Sales Meetings	( )	( )
g. Safety/Health	( )	( )	p. Product Demonstration	( )	( )
h. Empl. Benefits	( )	( )	q. Point-of-Sale	( )	( )
i. Other (pls spec.) _____	( )	( )	r. Annual Rpts/Meetings	( )	( )
			s. Sec. Analyst Presentations	( )	( )
			t. Community Relations	( )	( )
			u. Labor/Govt. Relations	( )	( )
			v. Other (pls specify) _____	( )	( )

33. Of the applications listed above, which are the **three most important** uses within your organization? (Rank in order by letter.) 1 ____ 2 ____ 3 ____

34. If another audiovisual medium is used for any of the above program applications **instead of video**, indicate the most important application by the appropriate letter.
    Slides ____ Film ____ Multi-media ____ Other _____ ____

35. How many programs were produced by your video operation in 1984?
    Communications Programs ____ Training Programs ____
    Other ____ Type? _____

36. How many will be produced this year? Communications ____ Training ____
    Other ____ Type? _____

37. How many do you anticipate will be produced in 1986?
    Communications Programs ____ Training Programs ____
    Other ____ Type? _____

38. There is a trend in some organizations to produce short, single topic "video memos" or messages that are done quickly with a minimum of editing, titles, music, graphics, etc. These "non-programs" are produced instead of or in addition to regular programs that have some or all of the traditional broadcast-style production values. Does this apply to your operation? If so, what percent of your production could be considered "non-program" video production of this type?
    1 ( ) produce only regular-style programs   2 ( ) produce both
    ____% are short video memos or messages

39. Do you ever transfer films and/or slide programs to videotape for distribution as programs?
    1 ( ) Often   2 ( ) Sometimes   3 ( ) Never

40. What percentage of your production is for the following functions in your organization:
    Top Management ____% Marketing/Sales ____% Legal ____%
    Corp. Communications ____% Line Divisions ____% Systems/EDP ____%
    Personnel/Training ____% Treasury/Financial ____% Other "Clients" ____%

41. What is the average length of the programs produced by your operation?
    1 ( ) less than 10 min.   3 ( ) 20 - 30 min.   5 ( ) 45 - 60 min.
    2 ( ) 10 - 20 min.   4 ( ) 30 - 45 min.   6 ( ) over 60 min.

42. Are your video programs produced using in-house people and equipment or through the use of outside services and facilities, or both?
    1 ( ) We do everything in-house.   2 ( ) Everything is done outside.
    3 ( ) We do some work in-house and some work outside.

43. If your programs are produced using both in-house facilities and outside services, what percentage of your programs use outside services for the following?
    Program concepts and scripting? ____%
    Principal production? ____%
    Off-line screening and editing? ____%
    Final editing/post production? ____%
    Duplication? ____%

44. What is the average cost of the programs your operation produces?
    1 ( ) under $2,000   4 ( ) $10,000 - 15,000   7 ( ) $25,001 - 35,000
    2 ( ) $2,000 - 5,000 5 ( ) $15,001 - 20,000   8 ( ) $35,001 - 50,000
    3 ( ) $5,001 - 10,000 6 ( ) $20,001 - 25,000   9 ( ) over $50,000

45. Does this average cost include ...
    1 ( ) All costs including staff salaries, overhead, out-of-pocket expenses, production costs, equipment, facilities, outside services, etc.?
    2 ( ) Only out-of-pocket, direct costs for each production?

46. Do you charge back any or all of your costs of program production and distribution to the unit or department requesting the program?
    1 ( ) Charge back all costs (including overhead and staff salaries).
    2 ( ) Charge back only materials, outside services and other direct costs.
    3 ( ) Charge back only duplication and/or distribution costs.
    4 ( ) Do not charge back any costs or expenses.

47. How many "published" programs will your video organization purchase this year? ____
    Indicate the three leading subject areas from the list in Question 33.
    1 ____ 2 ____ 3 ____

48. How involved is your video operation in the selection and purchase of these programs?
    1 ( ) Do the selection and buying.   2 ( ) Help select and/or help buy.
    3 ( ) Not involved in either selection or purchase of published programs.

Survey Questionnaire 161

PRODUCTION

49. How many of the following types of equipment do you have in your operation?
    ENG cameras _____ Off-line editing units _____ Duplicating units _____
    Studio cameras _____ Post-production editing _____ Playback units _____

50. If you have one, what is the approximate size of your in-house fixed studio or production area?
    1 ( ) No fixed studio/production area   3 ( ) 600 - 1,250 sq.ft.
    2 ( ) Under 600 sq. ft.   4 ( ) Over 1,250 sq. ft.

51. Where is this facility located?
    1 ( ) No fixed studio/production area   4 ( ) Organization hdqrtrs
    2 ( ) Training center (not in hdqtrs)   5 ( ) Other _____
    3 ( ) A/V media center (not in hdqtrs)   (pls specify)

52. Of the programs produced this year, what percent will be shot on location as opposed being shot in a studio (either in-house or hired)? _____%

53. How many of your programs will be shot in any of the following formats:
    1/2" VHS _____   1/2" Betacam _____   3/4" U-Mat. _____   2" _____
    1/2" Beta _____ 1/2" "M" Format _____ 1" type C _____   Other _____

54. On which of these formats are most of your programs mastered? (Indicate by number from Question 53 above.) # _____

55. Are you using computer-generated graphics in the production of slides, print graphics and/or video graphics?
    ( ) Slides   ( ) Print graphics   ( ) Graphics for video

56. Does your operation have the need to produce electronically-generated graphics for high quality print reproduction (Annual Reports, brochures, etc.)? ( ) Yes ( ) No

57. If yes, what is the most your organization would be willing to pay for a system that would provide this capability? $_____

58. A machine which will produce high quality (but not reproduction print quality) 5" x 6" color prints of any video frame and some computer displays will soon be introduced. Would such a machine be useful in your video operation? ( ) Yes ( ) No

59. If yes, what would it be used for? _____

60. What would be the most you would be willing to pay for such a unit? $_____

Several new systems are being developed around expanded versions of standard desk-top personal computers which will perform a number of video production tasks in addition to running standard word processing, spreadsheet, database, project management and other business programs. The video production capabilities include digitalizing the video input and providing a variety of special effects manipulations of the video image (similar to the Quantel and Dubner systems). In addition, the systems have a virtually unlimited character generator and computer/video graphics capability and can provide some animation. The systems also function as top-of-the-line post-production and editing computers that will interface with any VTR. Picture quality is equivalent to 3/4 in. broadcast, making it suitable for most corporate/institutional video applications.

61. How interested would you be in such a system for your video operation?
    1 ( ) Highly interested   3 ( ) Possibly interested
    2 ( ) Interested   4 ( ) Not interested

62. What is the most your organization would be willing to pay for such a system?
    $_____.

63. Check which applications you would use such a system for.
    a. ( ) Char./Graphics generator   d. ( ) Scriptwriting/word processing
    b. ( ) Digital special effects   e. ( ) Budgeting, estimating, cost control
    c. ( ) Editing/Post-production   f. ( ) Project management
         g. ( ) Other _____

64. Which would be the three most important applications to you? (Rank by letter)
    1 _____   2 _____   3 _____

Several manufacturers have recently introduced to the consumer market a new 8 mm camera/recorder combination the size of a small book that features a video cassette similar to an audio cassette. Picture quality is said to be comparable to either of the 1/2-in. formats. The format has been standardized and is supported by many different manufacturers. Including a separate recorder/player unit and accessories, the system is priced under $1,800.

65. In what way do you see such a system being used in your organization?
 1 ( ) See no use in our organization
 2 ( ) Would use for some production in our video operation
 3 ( ) Would be used by field units to produce own video programs
 4 ( ) Both #2 and #3
 5 ( ) Other _____

66. If you see no use for such a system in your organization, please indicate why.
 _____

67. If you could use the new 8 mm system in your own video production, please tell how.
 _____

68. If you see the new 8 mm system being used by field units, what applications would they use it for?
 1 ( ) Point-of-sale   4 ( ) Field-to-headquarters communications
 2 ( ) Product demonstration 5 ( ) Personal development/role playing
 3 ( ) Local communications/ 6 ( ) Job/skill training
       Employee information   7 ( ) Other _____

69. If the new 8 mm system were to be used by field units, what help would you provide them?
 1 ( ) Equip. selection/purchase   5 ( ) Duplication
 2 ( ) Video production training   6 ( ) Other _____
 3 ( ) Conversion to other formats
 4 ( ) Post prod./editing services 7 ( ) Would provide no assistance

70. Would videotapes produced by individual field units be viewed only within that unit, distributed to other locations or both?
 1 ( ) Viewed only in that unit   3 ( ) Both
 2 ( ) Sent to other locations   4 ( ) Don't know

71. The 8mm tape system also is capable of recording up to 24 hours of high-quality digital audio. Would this feature be useful in your organization? If so, how?
 ( ) Yes   ( ) No   If yes, how? _____

DISTRIBUTION

72. What is the total number of locations in your organization where video programs can be viewed?
 1 ( ) Less than 5   5 ( ) 36 - 50   9 ( ) 301 - 400
 2 ( ) 5 - 10   6 ( ) 51 - 100   10 ( ) 401 - 500
 3 ( ) 11 - 20   7 ( ) 101 - 200 11 ( ) 501 - 1000
 4 ( ) 21 - 35   8 ( ) 201 - 300 12 ( ) Over 1000

73. Roughly how many locations would have more than one playback unit? _____

74. How many of these are outside your country? _____

75. Are each of your programs sent to all locations? If not, on the average to how many locations does each program go? ( ) Send to all locations
 1 ( ) Less than 5   5 ( ) 36 - 50   9 ( ) 301 - 400
 2 ( ) 5 - 10   6 ( ) 51 - 100   10 ( ) 401 - 500
 3 ( ) 11 - 20   7 ( ) 101 - 200 11 ( ) 501 - 1000
 4 ( ) 21 - 35   8 ( ) 201 - 300 12 ( ) Over 1000

76. How many do you distribute outside your country? _____

77. Do you, or are you planning to, distribute your programs in more than one language? If so, how many?
 (1) Only one language   3 ( ) 3-4 languages
 (2) 2 languages   4 ( ) More than four languages

78. On the average, how many copies are made of each program?
 1 ( ) Less than 5   5 ( ) 36 - 50   9 ( ) 301 - 400
 2 ( ) 5 - 10   6 ( ) 51 - 100   10 ( ) 401 - 500
 3 ( ) 11 - 20   7 ( ) 101 - 200 11 ( ) 501 - 1000
 4 ( ) 21 - 35   8 ( ) 201 - 300 12 ( ) Over 1000

Survey Questionnaire

79. Once a program is completed, how long does it usually take to have duplicated and distributed?
    1 ( ) 24 hours or less   3 ( ) 2 - 5 days
    2 ( ) 24 - 48 hours   4 ( ) 1 - 2 weeks

80. Will your distribution network be expanded or decreased in the next year? If so, by about what percent?
    ( ) Network will NOT expand   ( ) Network will expand ____%   ( ) Network will decrease ____%

81. In general, how are your programs viewed?
    1 ( ) Individually, at viewer's discretion
    2 ( ) Small group sessions (less than 10 people)
    3 ( ) Large group sessions (more than 10 people)
    4 ( ) At customer or point-of-sale locations
    5 ( ) Outside group meetings
    6 ( ) Don't know ... send out for field to use as they see fit

82. How many times do programs tend to be played or viewed in most locations?
    1 ( ) once   2 ( ) 2 - 3 times   3 ( ) 4 - 10 times   4 ( ) more than 10 times

83. Do you try to measure audience reaction or validate the results of the programs you distribute?
    1 ( ) Training programs only   3 ( ) All programs
    2 ( ) Communications programs only 4 ( ) Do no measurement or validation

84. If yes, what type of validation or measurement do you use?
    1 ( ) Pre-test/post-test   5 ( ) Phone surveys
    2 ( ) Questionnaires/feedback forms 6 ( ) Q & A sessions following viewing
    3 ( ) Informal comments   7 ( ) Viewing logs
    4 ( ) Random interviews   8 ( ) Data from interactive systems
              9 ( ) Other _____

85. Do you use any form of promotion to stimulate viewing of your programs? (More than one may apply)
    1 ( ) Posters   4 ( ) Bulletin Bds.   7 ( ) Telephone
    2 ( ) Fliers   5 ( ) Catalogues   8 ( ) Other _____
    3 ( ) Empl. Publ.   6 ( ) Mail/memos   9 ( ) Do not promote viewing

86. Check any of the following methods of program distribution that you use.
    1 ( ) 3/4" videocassette   4 ( ) Other videocassette   7 ( ) Microwave
    2 ( ) 1/2" VHS cassette   5 ( ) Videodisc   8 ( ) Satellite
    3 ( ) 1/2" Beta cassette   6 ( ) In-house cable   9 ( ) Other _____

87. Which of the above is the primary means of distribution for your programs? No. _____

88. Do you use outside duplication services for most of your programs? ( ) Yes ( ) No

89. Do you see any problem in distributing programs in more than one videotape format? If so, why? ( ) No ( ) Yes
    _____

90. Approximately how many videotapes in the following sizes and formats will your operation purchase this year?

VHS	Beta	U-Matic	"C" 1-in.
T10/20 ___	L125 ___	BU5 ___ U10 ___	34 ___
T30 ___	L250 ___	BU10 ___ U20 ___	66 ___
T60 ___	L370 ___	BU18 ___ U30 ___	96 ___
T90 ___	L500 ___	BU30 ___ U45 ___	125 ___
T120 ___	L750 ___	BU45 ___ U60 ___	155 ___
PV10 ___	PB10 ___	BU60 ___	188 ___
PV20 ___	PB20 ___		

91. What is the primary brand purchased in each format?
    VHS _____ BETA _____ U-MATIC _____ "C" 1-in. _____

92. Do you re-cycle videotapes so they can be used again for distribution? If so, approximately how many times? ( ) No ( ) Yes _____

93. When buying videotapes, what influences your selection the most ...
    1 ( ) Name brand   3 ( ) Experience   5 ( ) Technical specifications
    2 ( ) Price   4 ( ) Availability 6 ( ) Recommendations of others

94. Are you involved in any aspect of office automation planning in your organization, particularly the use of "Local Area Networks" (LANs)?
    ( ) No   ( ) Yes   How? _____

95. If a LAN were installed in your organization which would permit the delivery of full-motion color video to every terminal workstation along with data and other forms of communications, how desireable would this be?
    1 ( ) Highly desireable   3 ( ) Possibly desireable
    2 ( ) Somewhat desireable   4 ( ) Not desireable

VIDEO PROJECTION

96. Does your organization use any form of video projection, either in the showing of video programs, in videoconferencing, or for data display?
    1 ( ) Showing programs   3 ( ) Data display
    2 ( ) Videoconferencing   4 ( ) Do not use video projection at all

97. If used to show programs, how often is it used?
    1 ( ) Regularly 2 ( ) Occasionally   3 ( ) Do not use video projection

98. If used to show programs, is it always at the same locations or do the locations vary?
    1 ( ) Always same location   2 ( ) Location varies

99. If the same location, please describe _____

100. Does your organization own the video projection equipment it uses, or does it rent the equipment as needed ... or both?
    1 ( ) Own equipment   2 ( ) Rent equipment 3 ( ) Both own and rent equipment

101. If you use video projection, what screen size is generally used?
    1 ( ) 4 ft. or less   3 ( ) 7 – 8 ft.   5 ( ) 12 ft.
    2 ( ) 5 – 6 ft.   4 ( ) 10 ft.   6 ( ) 14 ft. or larger

102. What is the average audience size when programs are shown using video projection equipment?
    1 ( ) less than 5   4 ( ) 16 – 20   7 ( ) 51 – 75   10 ( ) 201 – 300
    2 ( ) 6 – 10   5 ( ) 21 – 30   8 ( ) 76 – 100   11 ( ) 301 – 400
    3 ( ) 11 – 15   6 ( ) 31 – 50   9 ( ) 101 – 200 12 ( ) over 400

103. Several manufactures have recently introduced small, easy-to-carry, light-weight, video projectors with built-in video cassette players, PA audio systems and other presentation features that can be set up anywhere quickly and are often priced under $3,000. Do you see a need or application for such a unit in your organization? If so, how many?
    1 ( ) Have no need   2 ( ) Could use ____ such units

104. Where and how would these portable video projectors be used?
    (Describe) _____

105. Would these portable units be used in place of or in addition to your conventional playback units?
    1 ( ) Instead of conventional units   2 ( ) In addition to conventional units
    3 ( ) Both

INTERACTIVE VIDEO

106. Does your organization now use, or plan to use, interactive video — that is, **interactive videotape, interactive videodisc** or **interactive teletext/videotex** — in any way?
    1 ( ) Now using   2 ( ) Planning to use   3 ( ) No plans to use interactive video

107. If now using or planning to use interactive video, which medium will be used?
    1 ( ) Videotape   2 ( ) Videodisc   3 ( ) Videotex   4 ( ) Combination ___ and ___

108. How many interactive video programs will your organization produce in 1985?
    1 ( ) 1 – 3   2 ( ) 4 – 6   3 ( ) 7 – 10   4 ( ) Over 10 No. _____

109. Will this number increase or decrease next year?
    1 ( ) Increase   2 ( ) Decrease   3 ( ) Will stay the same

Survey Questionnaire

110. For what applications?
     ( ) Job/skill training   ( ) Sales/marketing   ( ) General information   ( ) Point-of-sale
111. In how many locations do you use interactive video?
     1 ( ) 1 - 3   2 ( ) 4 - 6   3 ( ) 7 - 10   4 ( ) Over 10   No. _____
112. Will this number increase or decrease next year?
     1 ( ) Increase   2 ( ) Decrease   3 ( ) Will stay the same
113. How many copies of each interactive video program are made?
     1 ( ) 1 - 3   2 ( ) 4 - 6   3 ( ) 7 - 10   4 ( ) Over 10   No. _____
114. Do you see the videodisc replacing videotape in certain applications, or is it used primarily for applications where tape was not previously employed?

     1 ( ) Replaces tape in some applications. Which? _____
     2 ( ) Used primarily for new applications. Which? _____
     3 ( ) Both
115. How involved is your video operation in the organization's use of interactive video?
     1 ( ) Responsible for all aspects from program design through production and replication
     2 ( ) Responsible for program design
     3 ( ) Responsible for programming/authoring
     4 ( ) Responsible for program production
     5 ( ) Combination _____ and _____
     6 ( ) Not involved in organization's use of interactive video
116. If you are not involved in program design, what department is? _____
117. If you are not involved in programming/authoring, what department is? _____
118. If you are not involved in production, what department is? _____
119. There is a new generation of interactive video equipment which combines a powerful desk-top computer with an optical laser video disc in a single unit. The system uses the disk's enormous capacity to store both data and video images which can be displayed separately or overlaid in virtually unlimited combinations. How desirable would such a system be in your organization?

     1 ( ) Highly desirable   2 ( ) Somewhat desirable   3 ( ) No need
120. There are new videotext workstations that can retrieve and display interactive training and information programs from a central computer in color with animated graphics. How desirable would such a system be in your organization?

     1 ( ) Highly desirable   2 ( ) Somewhat desirable   3 ( ) No need

PROFESSIONAL EDUCATION & TRAINING

121. What is your highest level of education?

     1 ( ) High School   3 ( ) Associates Degree   5 ( ) Masters/Prof. Degree
     2 ( ) Some College   4 ( ) Bachelors Degree   6 ( ) Doctorate
122. If you attended college, was your major in ...

     1 ( ) Non-bdcst media   4 ( ) Speech/Communications   7 ( ) Business
     2 ( ) Broadcasting   5 ( ) English   8 ( ) Eng./Technical
     3 ( ) Journalism   6 ( ) Liberal Arts   9 ( ) Education
                            10 ( ) Other _____
123. Have you continued your media-related training in the last five years by ... (Check all that apply)

     1 ( ) Attending conferences   4 ( ) Reading professional books/journals
     2 ( ) Coll./univ. courses   5 ( ) In-house workshops using outside experts
     3 ( ) Special training courses   6 ( ) Other _____

124. Which one of these areas has been of most importance in your career development?

    ( ) Media Production   ( ) Educ./Instr. design   ( ) Advertising/Marketing
    ( ) Script writing   ( ) Psychol./Hum.learning ( ) Engineering
    ( ) Management   ( ) Other _____

125. Please rank the following areas in order of priority for you and your staff's training.

    Media production _____   Computers _____
    Writing/design _____   New media/technologies _____
    Interactive video _____   Engineering/maintenance _____
    Management _____   Other _____

126. Does your department have a budget for continuing professional education or training? If so, how much is allocated each year?

    1 ( ) Under $500   3 ( ) $1,001 - 2,000   5 ( ) $3,001 - 5,000
    2 ( ) $501 - 1,000   4 ( ) $2,001 - 3,000   6 ( ) Over $5,000

127. What career path do you perceive yourself following within your organization?

    1 ( ) Remaining a video or audiovisual professional
    2 ( ) Moving into management and broader areas of responsibility
    3 ( ) Leaving the organization in order to pursue personal goals

128. How supportive is your organization in furthering your professional development?

    1 ( ) Very supportive   2 ( ) Somewhat supportive   3 ( ) Not supportive

MANY THANKS FOR YOUR HELP AND COOPERATION. PLEASE RETURN THE QUESTIONNAIRE IN THE ENCLOSED POSTAGE PAID ENVELOPE BY SEPTEMBER 11.

Survey Questionnaire   167

PRIVATE TELEVISION COMMUNICATIONS CONFIDENTIAL QUESTIONNAIRE: 1985
--- PART TWO: V I D E O C O N F E R E N C I N G ---

If your organization is not using videoconferencing in any way and has no plans to do so, discard this section. If your organization is using, or is planning to use, videoconferencing in the foreseeable future and you will be directly involved, please continue with **Question Number 129** below. If someone else in your organization is more directly involved with videoconferencing, please have them fill out this _entire_ section, including Questions 1 and 2. Thank you.

Organization Name: _____
Your Name & Title: _____

1. Primary business/organization classification:

   1 ( ) Manufacturing   10 ( ) Wholesale/Retail
   2 ( ) Natural Res.(petrl/mining/frst.prod.) 11 ( ) Travel & Lodging
   3 ( ) Construction/Real Estate   12 ( ) Publishing/Entertainment
   4 ( ) Communications/Electronics   13 ( ) Service Industry
   5 ( ) Insurance   14 ( ) Conglomerate
   6 ( ) Banking/Other Financial   15 ( ) Medical Services
   7 ( ) Utility (telephone,elec.,gas)   16 ( ) Education
   8 ( ) Transportation/Shipping   17 ( ) Government
   9 ( ) Food/Beverage   18 ( ) Other _____

---

129. Is your organization now using any form of satellite videoconferencing? ( ) Yes ( ) No

130. If not, has it used satellite videoconferencing in the past? ( ) Yes ( ) No

131. Were these one-way, point-to-multipoint, _ad hoc_ videoconferences or two-way, point-to-point conferences?
     1 ( ) One-way, ad hoc   2 ( ) Two-way   3 ( ) Have used _both_ forms

132. How many videoconferences has your organization held in the last two years?
     1 ( ) One-way _____   2 ( ) Two-way _____

If your organization has not used satellite videoconferencing to date...

133. Do you know of any plans to use it in the near future? If so, how soon?
     ( ) Will not be using soon   ( ) Will be using soon   When? _____

If your organization has used videoconferencing ...

134. Where does the idea for doing a videoconference generally originate?
     1 ( ) Within the video operation   2 ( ) Another department Who? _____

135. Who is generally responsible for the planning and production of a videoconference, including selection of the network vendor?
     1 ( ) Video operation   2 ( ) Another department Who? _____

136. If your video operation is **not** responsible for the planning and production of videoconferences, do you assist the department that is responsible in any way?
     1 ( ) Assist in planning/production   2 ( ) Usually not involved

137. What was the average number of locations involved in the one-way conferences? _____

138. If the number of locations varies for each videoconference, what is the largest number of locations that have been involved for any of these one-way conferences? _____ The smallest number of locations? _____

139. In general, are most of these receiving locations at company facilities or at rented off-site locations?
     1 ( ) Company facilites   2 ( ) Off-site locations   3 ( ) Combination

140. If a combination, how many company facilities and how many off-site facilities are involved?
     No. company locations _____   No. off-site locations _____

141. On the average, how many people take part in these one-way videoconferences per location?
     1 ( ) Less than 3   3 ( ) 5 - 10   5 ( ) 21 - 50   7 ( ) 101 - 300
     2 ( ) 3-5   4 ( ) 11 - 20   6 ( ) 51 - 100   8 ( ) Over 300

142. What level of managers/employees are generally involved at these locations?
    1 ( ) Sr. Management   4 ( ) Engineering/Technical
    2 ( ) Middle Management   5 ( ) Staff Specialists
    3 ( ) 1st Line Superv.   6 ( ) Other

143. For which of the following purposes were the one-way conferences used?
    1 ( ) Sales/Marketing   4 ( ) Training/Education
    2 ( ) Public Relations   5 ( ) Engineering/ Prod. Planning
    3 ( ) Mgt/Employee Infor   6 ( ) Other _____

144. Have any of these one-way videoconferences been directed to audiences outside your organization? ( ) Yes ( ) No

145. Does your organization have a continuing need to reach outside audiences with one-way videoconferencing? ( ) Yes ( ) No

146. Do one-way video conferences originate in your own in-house facilities or at an outside facility?
    1 ( ) In-house facility   2 ( ) Outside facility

147. Was two-way audio used with any of these videoconferences?
    1 ( ) Used two-way audio with all   2 ( ) Only some   3 ( ) None

148. Was video projection used with any of these videoconferences?
    1 ( ) Video projection used with all   2 ( ) Used both monitors and video projection
    3 ( ) Used no video projection, monitors only

149. In general, how satisfied is the audience with the quality of communications provided by one-way videoconferencing?
    1 ( ) Very satisfied   3 ( ) Somewhat dissatisfied
    2 ( ) Somewhat satisfied   4 ( ) Very dissatisfied

150. If there is dissatisfaction, what is the most commonly voiced complaint about one-way videoconferencing? _____

151. What do people like most about one-way videoconferencing? _____

152. Does your organization use or intend to use one-way videoconferences on a regular basis?
    ( ) Yes   ( ) No

153. If yes, do you now have, or intend to install, a permanent in-house origination facility?
    1 ( ) Now have in-house facility   2 ( ) Will install an in-house facility

154. If your organization intends to use one-way videoconferences on a regular basis, do you now have, or intend to install, permanent receiving equipment at company facilities? If so, at how many locations?
    Now have _____ permanent receiving locations
    Will install _____ permanent receiving locations

155. What is your estimated cost to set up and equip a permanent one-way receiving location, including earth station and related costs?
    1 ( ) Less than $3000   2 ( ) $3000 - 5000   3 ( ) $5000 - 7,000   4 ( ) Over $7,000

156. If your organization chose to set up a permanent, dedicated satellite videoconference network, would you prefer to purchase or lease the equipment?
    1 ( ) Prefer to own equipment   2 ( ) Prefer to lease equipment

157. If your organization has **not used** two-way videoconferencing, are there any plans to do so within the next two years? ( ) Yes ( ) No

158. If your organization **has used** two-way, point-to-point videoconferences, what purposes were they used for?
    1 ( ) Sales/Marketing   4 ( ) Training   7 ( ) Recruiting/Interviewing
    2 ( ) Public Relations   5 ( ) Engineering/ Prod. Planning   8 ( ) Other _____
    3 ( ) Mgt/Employee Infor   6 ( ) Mgt. Review/Planning Mtgs

159. What level of managers/employees are generally involved in two-way videoconferences?
    1 ( ) Sr. Management   4 ( ) Engineering/Technical
    2 ( ) Middle Management   5 ( ) Staff Specialists
    3 ( ) 1st Line Superv.   6 ( ) Other _____

160. Does your organization have its own installed two-way videoconferencing facilities?
    ( ) Yes ( ) No   How many locations are involved?
    1 ( ) Less than 3   2 ( ) 4 - 6   3 ( ) 7 - 10   4 ( ) Over 10

161. Approximately how many times a month is your two-way videoconferencing system used?
    1 ( ) Less than 5   2 ( ) 6 - 10   3 ( ) 11 - 20   4 ( ) Over 20 No. ____

APPENDIX B

Job Descriptions

Following is a rather comprehensive set of job descriptions covering a variety of levels and classifications of functions within the audiovisual area.

They are based on a hypothetical Audiovisual Services Department made up of two separate groups: a *Production* group consisting of engineering and technical talent; and, a *Program* group, made up of software or creative programming staffers. This organization is for illustration only and is not necessarily the way an Audiovisual Services or video operation must be structured.

We have given you reporting relationships as well as job responsibilities so that you can model your own job descriptions and/or organizational charts as you so choose.

MANAGER, AUDIOVISUAL SERVICES

Reports to: Corporate Communications, Director

Basic Function:

Responsible for planning, organizing, coordinating, staffing and budgeting all aspects of audiovisual media production including video, motion pictures, slide and multi-media presentations, audio-tape and filmstrips.

Principal Responsibilities:

1. Develop, recommend and coordinate policies and plans for application of existing and new video and audiovisual technology on an organization-wide basis.

2. Long-range and annual planning of organization's audiovisual programs involving the setting of specific program objectives in support of organizational goals and the determination of audiovisual capital and operating budgets.

3. Supervise the design and content of all materials produced by audiovisual services for both internal and external distribution.

4. Set and maintain the quality standards for all media systems.

5. Provide consultation to in-house client-users on the effective use of all media systems.

6. Oversee the scheduling, budgeting and overall production of all media projects.

7. Set up and supervise the operation of effective cost accounting systems.

8. Set up and maintain feedback and effectiveness measurement and program evaluation systems.

9. Set staffing requirements and supervise and evaluate the professional performance of all division personnel.

10. Supervise the training of all audiovisual personnel.

11. Establish and maintain the division's relationship with Corporate Communications and with the Audiovisual Policy Committee.

AUDIOVISUAL ADMINISTRATOR

Reports to: Manager, Audiovisual Services

Principal Responsibilities:

1. Maintain budget for Audiovisual Services, including assisting in the preparation of individual project budgets.

2. Develop and maintain a cost-accounting/cost-analysis system and develop a set of guidelines for distribution to client-users.

3. Prepare production budget accounts (invoices) for client-users.

4. Coordinate overall scheduling of facilities and staff.

5. Work with A/V Services management in establishing annual operating and capital budgets including staff, equipment, facilities and supplies.

6. Keep records on feedback and effectiveness measurement systems and report to client-users.

7. Coordinate distribution of completed programs to or by client-users and the return of these programs for librarying and/or reuse of the videotape.

8. Establish and maintain library of generic videotape and film production footage, slides, graphics, sound/music, etc. for future use; maintain library of edited masters and distribution copies as well as completed films; and maintain up-to-date inventory records on all audiovisual and video production and distribution equipment.

9. Develop and maintain computer program for the library, scheduling, budgeting and cost-accounting systems.

MANAGER, MEDIA SERVICES

Reports to: Manager, Audiovisual Services

Basic Function:

Supervision of the planning and production of all video and audiovisual programs in coordination with Production Services and/or outside production sources and client-users, including pre-production, production and post-production activities.

Principal Responsibilities:

1. Meet communications and training objectives by overseeing the production of professional and cost-effective programs and materials.

2. Provide consultation to client-users on the need for an effective use of audiovisual and video media, including setting program objectives and budgets, media selection and audience analysis.

3. Supervise and coordinate the content, creative form and technical aspects of all production through the activities of the Media Producers, maintaining the highest quality standards possible in keeping with agreed upon objectives and budgets.

4. Supervise, train and evaluate the performance of the Media Producers.

5. Implement a program feedback and effectiveness measurement system for all media.

6. Assist in developing the operating and capital budgets, including specification of staffing, equipment, facilities and supplies.

Job Descriptions 173

MEDIA PRODUCER

Reports to: Media Services Manager

Basic Function:

Coordinates all aspects of assigned audiovisual projects, including interpreting client-user needs, working with content specialists and/or writers, determining budget, supervising production with technical crew and delivering completed job to the client.

Principal Responsibilities:

1. Client-user contact on a continuing basis.

2. Interpret a client-user needs.

3. Help client-user determine program content, visualization and objectives.

4. Establish media to be used and the appropriate production budget.

5. Work with internal writers and/or content specialists; secure outside specialist services when needed or requested or work with outside specialists contracted by client-user for program content development.

6. Manage all aspects of assigned productions, including working with graphic designers, set designers, technical crew and other participants in the project's production.

7. Audition talent to appear on camera or assist in their selection from internal and/or external sources.

8. See that the program is completed on time, within budget and to the client-user's satisfaction.

9. Assist in evaluating program results with client-users.

10. Participate in evaluation of all programs and media as projects are completed by Media Services unit.

PRODUCTION COORDINATOR

Reports to: Media Services Manager

Basic Function:

Assist Media Services Manager and/or Media Producers in coordinating efforts of the Media Services unit.

Principal Responsibilities:

1. Assist in scheduling and budgeting of all audiovisual projects undertaken by Media Services unit.

2. Keep track of individual project budgets and prepare reports to be submitted to client-users by Audiovisual Services.

3. Assist Media Producers in completing projects by acting as script person, coordinating sets and props, working with graphics and other support services, setting pre-production meetings, calling auditions and rehearsals, etc.

Job Descriptions 175

MANAGER, PRODUCTION SERVICES

Reports to: Manager, Audiovisual Services

Basic Function:

Manage the operation of an in-house video and audiovisual production center.

Principal Responsibilities:

1. Supervise and coordinate the technical aspects of all forms of video and audiovisual production, maintaining the highest possible quality standards.

2. Supervise the acquisition and maintenance of all production equipment and keep abreast of technological developments and equipment in various segments of the video and audiovisual industries.

3. Provide consultation to Media Services and client-users on production techniques and the most effective use of production equipment.

4. Manage the scheduling, budgeting and overall production of all programs using in-house facilities, production personnel and equipment.

5. Develop an effective cost accounting/cost analysis system in coordination with the Audiovisual Administrator with the objective of preparing/maintaining a production rate card competitive with comparable outside services.

6. Develop annual operating and capital budgets in coordination with the Audiovisual Administrator, including specifications on staffing, equipment, facilities and supplies.

7. Supervise the training, professional development, performance and evaluation of the engineering and production staffs.

SUPERVISOR, PRODUCTION

Reports to: Manager, Production Services

Basic Function:

Supervise the production of all in-house audiovisual, film and video programs on location or in the studio.

Principal Responsibilities:

1. Work directly with the Media Producer and client users during pre-production planning as a consultant on various aspects of production; supervise and coordinate the actual production of programs; supervise the post-production and completion stages.

2. Direct the activities of the crews on location or in the studio.

3. Assist the Media Producer in preparing production budgets and help insure that programs are completed on time, within budget and to the user's satisfaction.

4. Maintain the highest possible video, film and audiovisual quality standards commensurate with the project's objectives and budget.

5. Supervise and evaluate the professional performance of all production personnel and oversee and encourage their professional development and training.

Job Descriptions 177

DIRECTOR/EDITOR

Reports to: Supervisor, Production

Basic Function:

Provide the professional expertise and creative judgment necessary to assemble a completed videotape, film or audiovisual presentation including both visual and audio elements.

Principal Responsibilities:

1. During the production of "live," multi-camera studio video programs, work under the supervision of the Media Producer directing the actions of various studio and control room personnel and making creative decisions on camera placement, composition and switching.

2. During editing, post-production and completion, handle assembly of all production elements into a finished master videotape or fine cut film.

3. Achieve and maintain top professional proficiency in the operation of all equipment associated with effects and editing, including production switchers, audio consoles, CMX and other automated editing systems.

4. Contribute creative and technical ideas and methods to help meet a project's objectives and budget.

CAMERA OPERATOR/LIGHTING DIRECTOR

Reports to: Supervisor, Production

Basic Function:

Design the lighting for studio and location productions and oversee the proper set up and operation of all camera equipment used in video, film and audiovisual production.

Principal Responsibilities:

1. Work with the Supervisor, Production and the program's Media Producer during pre-production to determine the proper lighting design and effects for the shoot.

2. Work with other camera operators and/or technicians to oversee the proper setting and balancing of all lights used in a production.

3. Oversee the set up and proper function of all camera equipment used in a production.

4. Work as camera operator during multi-camera production and as camera operator, as assigned, on single-camera productions.

5. During production, provide the program director with well-composed, suitably framed and sharply focused images and smoothly executed pans, tilts and zooms.

PRODUCTION TECHNICIAN

Reports to: Supervisor, Production

Basic Function:

Installation of sets and props and general assistance in both studio and location productions.

Principal Responsibilities:

1. Work with the program director on the design and installation of sets and props.

2. Move and position equipment and props as required during a production.

3. Assist the camera operators and lighting director in the setting of lights.

4. Handle the appropriate transportation of equipment and supplies for a location production.

SUPERVISOR, ENGINEERING

Reports to: Manager, Production Services

Basic Function:

Responsible for the technical performance of all in-house video, film and audio production equipment and facilities and the supervision of the technical personnel operating and maintaining this equipment.

Principal Responsibilities:

1. Supervise the proper set up and operation of all audiovisual equipment for both studio and remote productions.

2. Maintain a close liaison and working relationship with the Supervisor of Production, providing technical advice and guidance during all phases of production and post-production.

3. Train production personnel in the proper use of equipment and systems.

4. Continually review and evaluate new *production* developments, techniques and equipment, recommending new equipment as appropriate.

5. Set up and administer an on-going maintenance program for all audiovisual equipment, insuring a state of readiness and reliability for all productions.

6. Select and maintain contact with outside maintenance and service contractors and supervise their work on in-house equipment.

7. Supervise and continuously up-date the technical proficiency of engineering staff personnel.

8. Assist in the development of the A/V Services operating and capital budgets and in the development of an effective cost analysis system.

VIDEO ENGINEER

Reports to: Supervisor, Engineering

Basic Function:

Operation and maintenance of all in-house videotape production and post-production equipment.

Principal Responsibilities:

1. Operate all types of electronic equipment and instruments.

2. Troubleshoot equipment problems and perform minor maintenance.

3. Set up equipment for studio and remote productions.

4. Serve as a member of the technical or production crews as required.

5. Keep abreast of technical developments and equipment in the audiovisual industry, particularly the video segment.

6. Oversee work of outside suppliers and maintenance contractors.

7. Order and maintain an adequate inventory of parts or repairs.

AUDIO ENGINEER

Reports to: Supervisor, Engineering

Basic Function:

Operation and maintenance of all in-house audio recording and duplicating equipment and facilities.

Principal Responsibilities:

1. Insure reliable performance of all in-house audio equipment and facilities.

2. Operate all types of audio recording and duplication equipment and instruments.

3. Maintain audio equipment and troubleshoot equipment problems.

4. Work with outside contractors on installation and maintenance as necessary.

5. Set up and arrange audio equipment for studio and remote video and film productions.

6. Serve as a member of the technical or production crews during video and film production.

7. Keep abreast of technical developments and equipment in the audiovisual industry, particularly as it relates to audio recording and the use of audio in video and film production.

8. Order and maintain an adequate inventory of parts for repairs.

VIDEO TECHNICIAN

Reports to: Supervisor, Engineering

Basic Function:

Operate in-house videotape production and post-production equipment.

Principal Responsibilities:

1. Operate all types of electronic equipment and instruments.

2. Set up equipment for studio and remote video and film productions.

3. Serve as a member of the technical or production crews as required.

4. Assist in troubleshooting equipment problems and in performing minor equipment maintenance functions.

AUDIO TECHNICIAN

Reports to: Supervisor, Engineering

Basic Function:

Operation and maintenance of all in-house recording and duplicating equipment and facilities.

Principal Responsibilities:

1. Operate all types of audio equipment and instruments.

2. Set up audio equipment for studio and remote video and film productions.

3. Serve as a member of the technical or production crews during video and film productions.

4. Assist in troubleshooting equipment problems and in performing minor equipment maintenance functions.

APPENDIX C

Bibliography

"A New Era For Management", *Business Week*, April 25, 1983.

"Beyond Unions", *Business Week*, July 8, 1985.

"Business Computing, A Special Bonus Section", *USA Today*, June 16, 1985.

Brush, Judith M. & Douglas P., *Private Television Communications: Into the Eighties (The THIRD Brush Report)*, 1981, available through H I Press, Inc., Cold Spring, NY.

Drucker, Peter F., "Out of the Depression Cycle", *Wall Street Journal*, January 9, 1985.

Garreau, Joel, *The Nine Nations of North America*, Boston, MA: Houghton Mifflin Company, 1981.

Gayeski, Diane M., *Corporate and Instructional Video*, Englewood Cliffs, NJ: Prentice-Hall, Inc., 1983.

Gayeski, Diane M., Williams, David V., *Interactive Media*, Englewood Cliffs, NJ: Prentice-Hall, Inc., 1985.

Gratch, Alon, "Tamed Rebels Make Good Managers", *New York Times*, February 10, 1985.

Kanter, Rosabeth Moss, *The Change Masters: Innovation and Entrepreneurship in the American Corporation*, New York: NY: Simon & Schuster, Inc., 1985 (paperback).

Kleinfield, N. R., "When Scandal Haunts the Corridors", *New York Times*, July 7, 1985.

Mueller, Robert K., "Renaissance Managers: A New Breed for Tomorrow's Electronic Office", *Today's Office*, March, 1985.

Naisbitt, John, *Megatrends*, New York, NY: Warner Books, Inc., 1982.

"New Corporate Elite", *Business Week*, January 21, 1985.

Piore, Michael J., Sabel, Charles F., *The Second Industrial Divide*, New York, NY: Basic Books, 1984.

Rosow, Jerome M., Edtr., *View From the Top: Establishing the Foundation for the Future of Business*, New York, NY: Facts on File, 1985.

Sambul, Nathan J., *The Handbook of Private Television*, New York, NY: McGraw-Hill, Inc., 1982.

Toffler, Alvin, *The Third Wave*, New York, NY: Morrow and Co., 1980.

Van Deusen, Richard E., *Practical AV/Video Budgeting*, White Plains, NY: Knowledge Industry Publications, Inc., 1984.

Van Nostran, William, *The Nonbroadcast Television Writer's Handbook*, White Plains, NY: Knowledge Industry Publications, Inc., 1983.

Watkin, Edward, "The CEO As Consummate Communicator", *Today's Office*, May, 1985.

APPENDIX D

Publications

Video publications abound covering both the private television and home video marketplaces. New entries come upon the scene, seemingly flourish for a short period of time and then drift off, never to be seen again. We offer you a mainstay list of publications which we find most useful in our work.

AV Video, 25550 Hawthorne Blvd., Suite 314, Torrance, CA 90505

Broadcast Management/Engineering (BME), 295 Madison Ave., New York, NY 10016

E-ITV Magazine, 295 Madison Ave., New York, NY 10016

Electronic Media, 740 Rush St., Chicago, IL 60611

International Television News, 6311 N. O'Connor Rd., Suite 110-LB51, Irving, TX 75039

Millimeter, 826 Broadway, New York, NY 10003

TV Digest/Consumer Electronics, 1836 Jefferson Pl., NW, Washington, DC 10036

TeleSpan Newsletter, 50 W. Palm St., Altadena, CA 91001

Television and Video Production, PO Box 109, Maclaren House, Scarbrook Rd., Croydon, Surrey CR9 1QH, England

Training, 731 Hennepin Ave., Minneapolis, MN 55403

Video Manager, 701 Westchester Ave., White Plains, NY 10604

Video Systems, 9221 Quivira Rd., Overland Park, KS 66212

Video Trade News, 51 Sugar Hollow Rd., Danbury, CT 06810

Videography, 50 W. 23rd St., New York, NY 10010

NOTE: A new magazine to be introduced in Fall, 1986 is *Corporate Video*, a bi-monthly publication of ITVA and Media Horizons, Inc. Editorial offices will be: 50 W. 23rd St., New York, NY 10010.